Padjadjaran Earth Dialogues: International Symposium on Geophysical Issues 2015

IOP Conference Series: Earth and Environmental Science
Volume 29

Bandung, Indonesia
8 – 10 June 2015

Editors:

Yudi Rosandi
Herbert M. Urbassek
Hiroaki Yamanaka

ISBN: 978-1-5108-1954-2
ISSN: 1755-1307

Printed from e-media with permission by:

Curran Associates, Inc.
57 Morehouse Lane
Red Hook, NY 12571

CURRAN ASSOCIATES INC. proceedings.com

Some format issues inherent in the e-media version may also appear in this print version.

Printed by Curran Associates, Inc. (2016)

For permission requests, please contact the Institute of Physics
at the address below.

Institute of Physics
Dirac House, Temple Back
Bristol BS1 6BE UK

Phone: 44 1 17 929 7481
Fax: 44 1 17 920 0979

techtracking@iop.org

Additional copies of this publication are available from:

Curran Associates, Inc.
57 Morehouse Lane
Red Hook, NY 12571 USA
Phone: 845-758-0400
Fax: 845-758-2634
Email: curran@proceedings.com
Web: www.proceedings.com

Table of contents

Volume 29

Padjadjaran Earth Dialogues: International Symposium on Geophysical Issues, PEDISGI
8–10 June 2015, Bandung, Indonesia

Accepted papers received: 24 November 2015
Published online: 19 January 2016

Preface

Papers

IOP Conf. Series: Earth and Environmental Science **29** (2016) 011001 doi:10.1088/1755-1315/29/1/011001

Preface

This issue of *IOP Conference Series: Earth and Environmental Science* contains selected papers presented at the *Padjadjaran Earth Dialogues: International Symposium on Geophysical Issues*, PEDISGI. The meeting was held from June 8 to 10, 2015, at the *Bale-Sawala* of Universitas Padjadjaran in Jatinangor, Indonesia. The PEDISGI is a symposium to accommodate communication between researchers, in particular geophysicists and related scientists, and to enable sharing of knowledge and research findings concerning local and global geophysical issues. The symposium was attended by 126 participants and 64 contributors from Indonesian universities and the neighbouring countries in four categories, viz. Theoretical and Computational Geophysics, Environmental Geophysics, Geophysical Explorations, and Geophysical Instrumentations and Methods.

The symposium was accompanied by a dialog, discussing a chosen topic regarding environmental and geological problems of relevance for the Indonesian archipelago and the surrounding regions. For this first event the topic was "*The formation of Bandung-Basin between myths and facts: Exemplary cultural, geological and geophysical study on the evolution of the earth surface*," presented by invited speakers and local experts. This activity was aimed at extending our knowledge on this particular subject, which may have global impact. This topic was augmented by theoretical background lectures on the earth's surface formation, presented by the invited speakers of the symposium.

The meeting would not have been successful without the assistance of the local organizing committee. We want to specially thank Irwan A. Dharmawan for managing the programme, Anggie Susilawati and Mia U. Hasanah for the conference administration, and Dini Fitriani for financial management. We also thank the National Geographic Indonesia for its support via the Business to Business Collaboration Program.

Yudi Rosandi
Herbert M. Urbassek
Hiroaki Yamanaka
Guest Editors

IOP Conf. Series: Earth and Environmental Science **29** (2016) 011001 doi:10.1088/1755-1315/29/1/011001

Peer review statement

All papers published in this volume of *IOP Conference Series: Earth and Environmental Science* have been peer reviewed through processes administered by the proceedings Editors. Reviews were conducted by expert referees to the professional and scientific standards expected of a proceedings journal published by IOP Publishing.

Reservoir Characterization around Geothermal Field, West Java, Indonesia Derived from 4-D Seismic Tomography

Rexha Verdhora Ry[1] and A D Nugraha[2]

[1]Master Program of Geophysical Engineering, Faculty of Mining and Petroleum Engineering, Institut Teknologi Bandung, Ganesha 10, Bandung 40132, Indonesia
[2]Global Geophysics Group, Faculty of Mining and Petroleum Engineering, Institut Teknologi Bandung, Ganesha 10, Bandung 40132, Indonesia

E-mail: rexha.vry@gmail.com

Abstract. Observation of micro-seismic events induced by intensive geothermal exploitation in a particular geothermal field, located in West Java region, Indonesia was used to detect the fracture and permeability zone. Using local monitoring seismometer network, tomographic inversions were conducted for the three-dimensional Vp, Vs, and Vp/Vs structure of the reservoir for January – December 2007, January – December 2008, and January – December 2009. First, hypocenters location was relocated using joint hypocenter determination (JHD) method in purpose to estimate best location. Then, seismic tomographic inversions were conducted using delay time tomography for dataset of every year respectively. The travel times passing through the three-dimensional velocity model were calculated using ray tracing pseudo-bending method. Norm and gradient damping were added to constrain blocks without ray and to produce smooth solution model. The inversion algorithm was developed in Matlab environment. Our tomographic inversion results from 3-years of observations indicate the presence of low Vp, low Vs, and low Vp/Vs ratio at depths of about 1 – 3 km below sea level. These features were interpreted may be related to steam-saturated rock in the reservoir area of this geothermal field. The locations of the reservoir area were supported by the data of well-trajectory, where the zones of high Vp/Vs were observed around the injection wells and the zones of low Vp/Vs were observed around the production wells. The extensive low Vp/Vs anomaly that occupies the reservoir is getting stronger during the 3-years study period. This is probably attributed to depletion of pore liquid water in the reservoir and replacement with steam. Continuous monitoring of Vp, Vs, and Vp/Vs is an effective tool for geothermal reservoir characterization and depletion monitoring and can potentially provide information in parts of the reservoir which have not been drilled.

1. Introduction

Micro-seismic monitoring in geothermal field provides insight into induced and naturally occurring stress changes. This technique maps active fault failure on shear zones, as well as fluid compression, e.g. [1,2,3]. Intensive geothermal exploitation at the considered geothermal field in West Java, Indonesia, induces micro-seismic events which are monitored by a local seismometer network. The reservoir in this geothermal field is the dominance of the water reservoir [4]. Using micro-seismic data monitored by this network, tomographic inversions were conducted for the three-dimensional Vp, Vs, and Vp/Vs structure of the reservoir. These velocity structures can be used to analyze the presence of water or stream in reservoir, e.g. [5,6,7]. We used micro-seismic data occurred during three years of observation from the period of January – December 2007, January – December 2008, and January – December 2009. We used dataset of every year respectively. Only data recorded by at least 3 stations were used. The first dataset which events occurred in 2007, are composed of 453 micro-seismic events and consists of 1842 P wave and S wave phases. The dataset of 2008 are composed of 1439 micro-seismic events and consists of 6663 P wave and S wave phases. Furthermore, the last dataset which events occurred in 2009, are composed of 535 micro-seismic events and consists of 2813 P wave and S wave phases. In purpose to image velocity structure of reservoir with good resolution, we used dimension of 1 x 1 x 0.5 km^3 for the parameterization model blocks.

First, micro-seismic hypocenters were relocated using joint hypocenter determination (JHD) method in purpose to estimate best location of sources. We used the best and updated 1-D velocity model from previous study [8] (figure 1). Then using these results, seismic tomographic inversions were conducted using delay time tomography for dataset of every year respectively. The travel times on the three-dimensional velocity model were calculated using ray tracing pseudo-bending method.

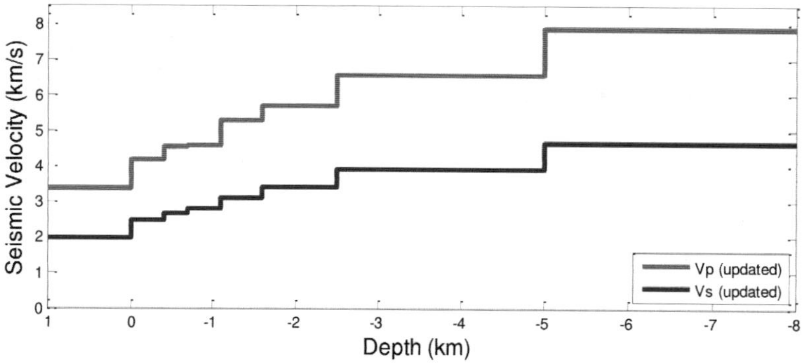

Figure 1. The 1-D velocity model used in JHD and as an initial model in tomographic inversion

The results of the processing conducted are three-dimensional velocity structures of Vp, Vs, and Vp/Vs for 2007, 2008, and 2009. These images of velocity structure may be able to characterize the condition and structure of reservoir. We significant found a strong Vp/Vs anomaly that correlates with the production zone. Theoretically, laboratory experiments and field studies suggest that this low Vp/Vs anomaly represents a zone where the pore fluid is predominately vapor, pressure is low, and the shear modulus is increased as a result of the drying of argillaceous material in the reservoir [9,10,11]. These changes probably resulted from the removal of reservoir fluids by production. The four-dimensional or time-lapse tomography can be used to study changes in reservoir structures with time. We detect a significant increment in the strength of low Vp/Vs anomaly. This may be attributed to progressive fluid depletion of the geothermal reservoir.

2. Methodology

2.1. Relocated Hypocenter

The precise 1-D velocity model and hypocenter location are the requirement as an input and initial model for seismic tomography. One of the methods to determine 1-D velocity model is coupled velocity-hypocenter method using Velest program [12]. The velocity models used were obtained using this method [8]. Then, we perform joint hypocenter determination (JHD) to solve the coupled hypocenter model problem for micro-seismic location and relocated 2427 micro-seismic events simultaneously.

2.2. Delay Time Tomography

The pseudo-bending method was used in this seismic tomography study as ray tracing method to determine possible ray path in 3-D velocity model and calculate synthetic travel time from hypocenter to receiver. Pseudo-bending [13] is an approach in minimization of travel time based on Fermat's Principle by giving small perturbations gradually on ray paths. This ray tracing algorithm in Matlab environment was developed by Syahputra *et al.* [14] and Fatkhan *et al.* [15].

In this study, we applied delay time tomography algorithm [16]. For resolve tomography inversion, least square method [17] was implemented iteratively to minimize the differences between observed and calculated travel times. We also added norm and gradient damping to constrain blocks without ray and to produce smooth solution model, respectively [16]. The inversion algorithm used was developed in Matlab environment.

3. Results and Discussion

Figure 2 shows the distribution of relocated micro-seismic events within study area. The range of hypocenters depth varies between elevations of 0.5 to -7 km. The deep events occurred at elevations of -5 to -6 km (mean sea level = 0 km) and the shallow events occurred at elevations -1 to 0.55 km. The distribution of these micro-seismic events in the east side of study area is well correlated to reservoir area. These events are shallow corresponded to permeability zone in reservoir. Meanwhile in the west side, the distribution of micro-seismic events is deeper than east side. These events correspond to the induced activities caused by injection well. In 2008, micro-seismic activities occurred almost three times greater than other years while these events occurred adjacent to injection well. High micro-seismicity in this area may indicate new fracture zone caused by fluid compression.

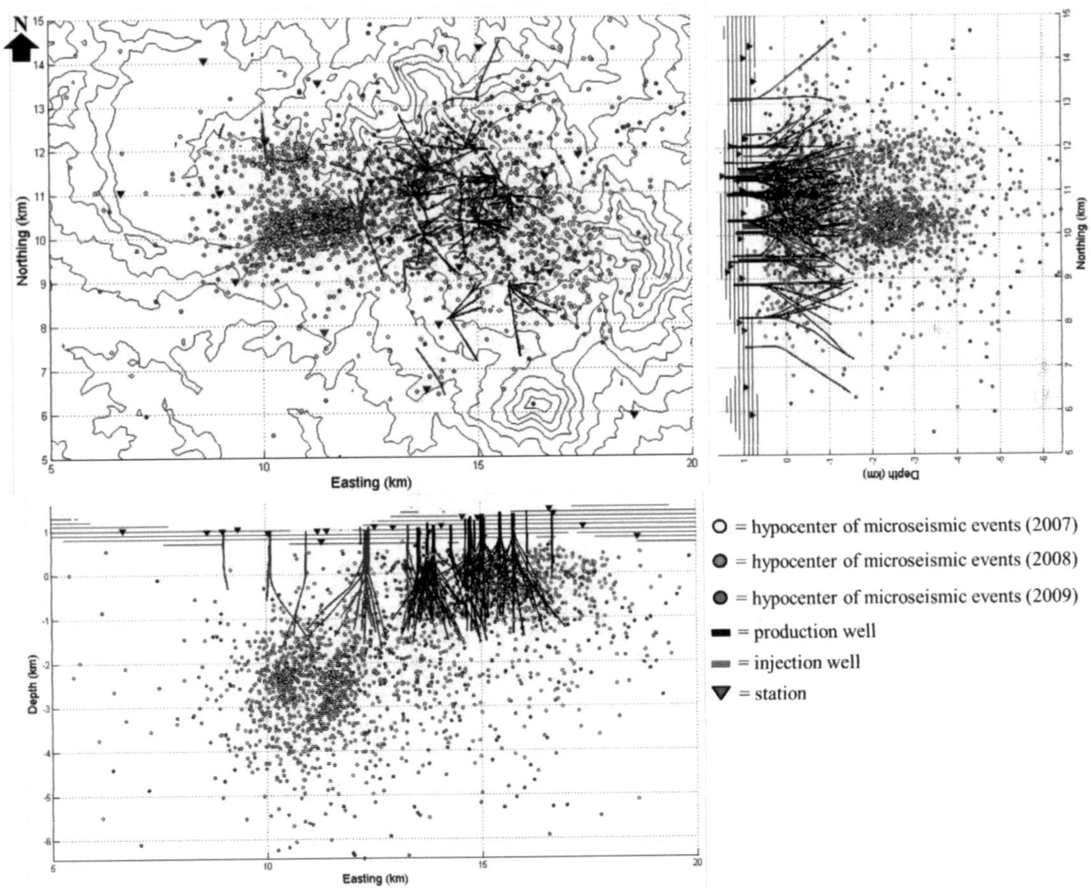

Figure 2. The map of the distributions of the relocated micro-earthquake events this study, well trajectory, and monitoring stations.

We then used those distributions of micro-seismic to image velocity structures around geothermal area for every year respectively. The parameterization model blocks used had dimensions of 1 x 1.5 x 0.5 km^3, so it can image velocity structure of reservoir with fine resolution. We also applied checkerboard resolution test (CRT) as a forward modeling method to test the reliability of the inversion technique used in tomographic inversion and to see the resolution throughout the model space. We used model anomalies ±10% of the initial 1-D velocity model. The results of CRT models for every year are shown at figure 3. The areas which have a pattern of checkerboard blue – red associate to the good resolution areas. We have the equal numbers of P and S wave phases in tomographic inversion, so the resolution for Vp and Vs are similar. Figure 4 shows the results of delay time tomographic inversion in depth section for Vp, while figure 5 shows the results for Vs. Only areas with good resolution are shown, corresponded to CRT.

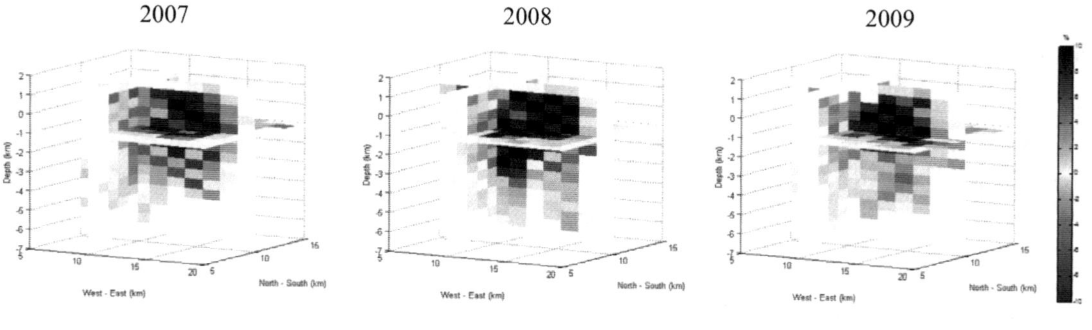

Figure 3. The 3-D structures of CRT model for 2007, 2008, and 2009.

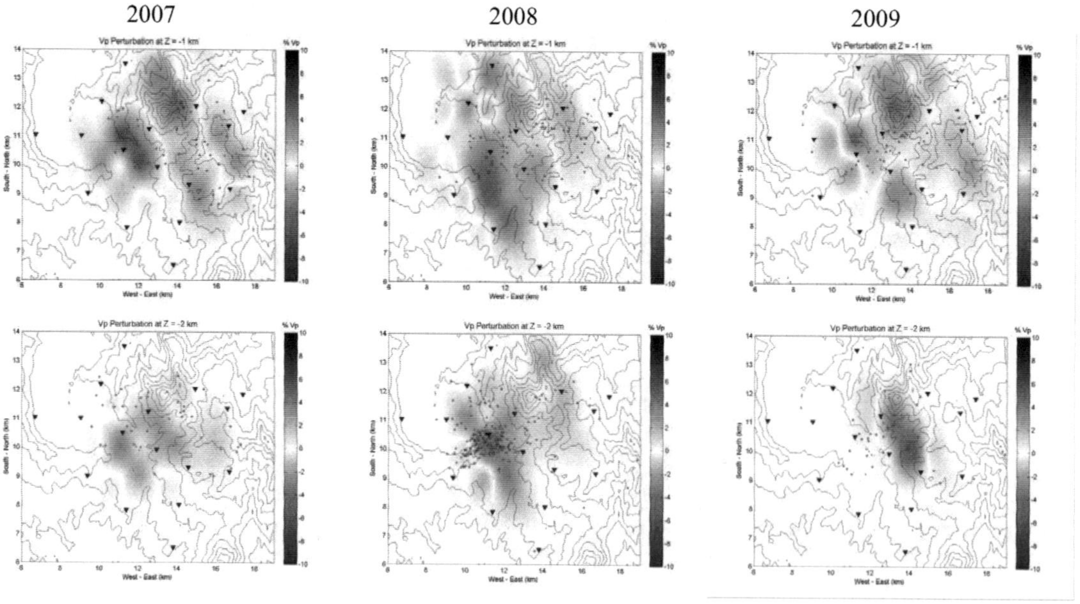

Figure 4. Tomogram of depth section for Vp anomaly in: 2007, 2008, and 2009 at elevation -1 and -2 km. Only areas with good resolution are shown. Positive anomalies are shown by blue colour and negative anomalies are shown by red colour.

Progressive and significant changes in the Vp and Vs fields occur with time. At elevation -1 km, anomaly of Vp decrease from 2007 to 2008, but increase in strength from 2008 to 2009. The anomaly of Vp at elevation -2 km has different pattern, where this anomaly decrease in strength every year from 2007 to 2009. Meanwhile for anomaly of Vs, at elevation -1 km, this anomaly increase in strength every year from 2007 to 2009. At elevation -2 km, the pattern is also different where anomaly of Vs decreases from 2007 to 2008, but increase in strength from 2008 to 2009. Factors that affect those changes include lithology, temperature, pore pressure, and pore fluid phase [11].

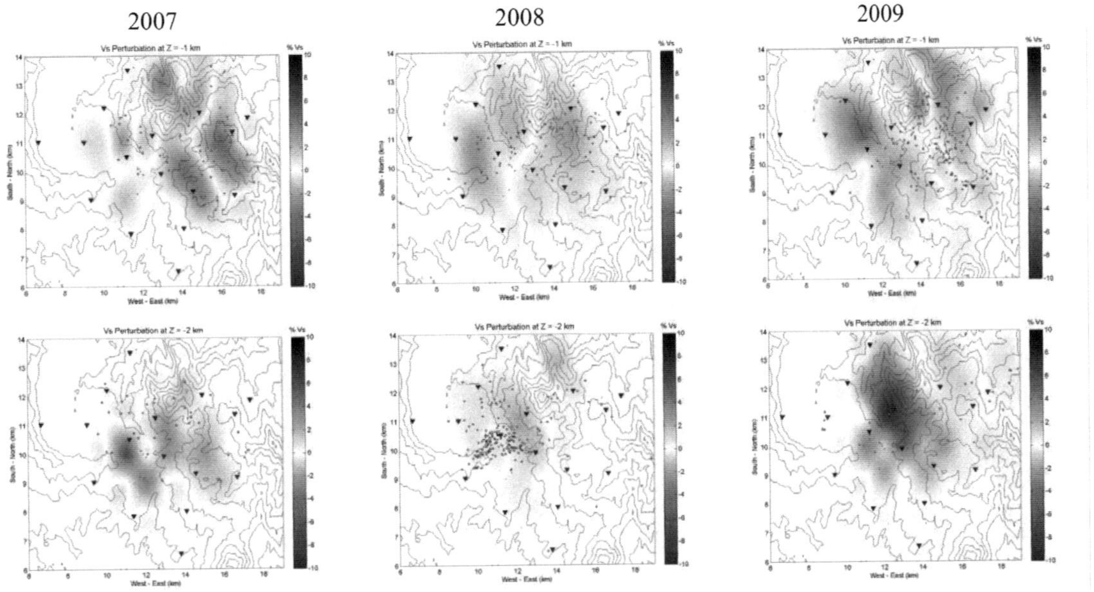

Figure 5. Tomogram of depth section for Vs anomaly in: 2007, 2008, and 2009 at elevation -1 and -2 km. Only areas with good resolution are shown. Positive anomalies are shown by blue colour and negative anomalies are shown by red colour.

Our interested area are shown in slice E–E' (or N-S 10 km) and F–F' (or N-S 11 km). Figure 6 shows the structures of Vp/Vs from 2007 until 2009 in this area. The zones of low Vp/Vs ratio (around 1.6) can be interpreted to be associated with steam-saturated rock, while the zones of high Vp/Vs (around 1.8) can be interpreted to be associated with water-saturated rock [9,10]. In general, the pattern of Vp/Vs anomaly growth is smooth with time. We identified those zones of low Vp/Vs as the reservoir of the considered geothermal field. The reservoir zones are located at 12 - 18 km WE, 8 - 12 km NS, and depth of 1 - 3 km below MSL. The existences of the reservoir area are supported by the data of well-trajectory. The pattern of Vp/Vs anomaly for every year is similar, where the zones of high Vp/Vs are around the injection wells and the zones of low Vp/Vs are around the production wells.

The reservoir in this geothermal field is the dominance of the water reservoir [4]. It becomes interesting when tomographic imaging from 2007 until 2009 identify the zones in reservoir area to be associated with steam-saturated rock. The exploitation and production of the geothermal field can cause changes in the reservoir phase system [11]. The progressive depletion of pore fluid causes the replacement of pore fluid with vapor. In addition, the pressure drop in the reservoir causes a decrease in the boiling point, resulting in boiling and vapor phase is formed.

Figure 7 shows clearly that the volume body of low Vp/Vs anomaly around reservoir area increase in strength fairly over 3-years period. Changes in pore pressure, pore fluid phase, and the water content of minerals caused by steam removal are probably responsible for the changes observed in the anomaly. The decrements of Vp, Vs, and Vp/Vs anomaly are attributed to steam flooding, while the increments of Vp, Vs anomaly and decrement of Vp/Vs anomaly are attributed to pressure change and drying of mineral [11]. Progressive growth in Vp/Vs ultimately results from depletion of liquid water in the reservoir; indeed different effects may be dominant in different parts of the field. The features at shallow part (elevation ~ -1km) are related to effect of pressure decrease and mineral drying. Meanwhile, the features at deeper part (elevation ~ -2km) are related to replacement of pore water by steam as the predominant effect.

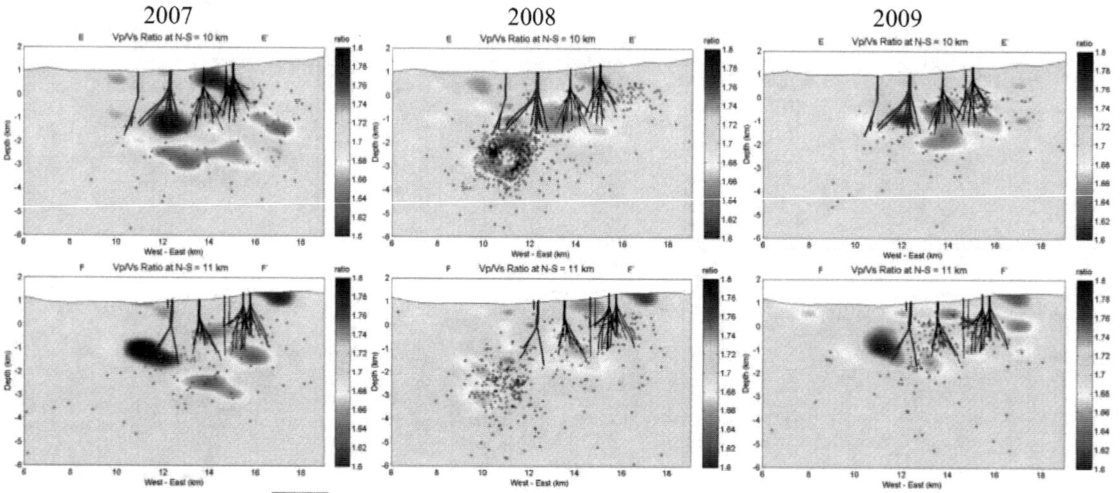

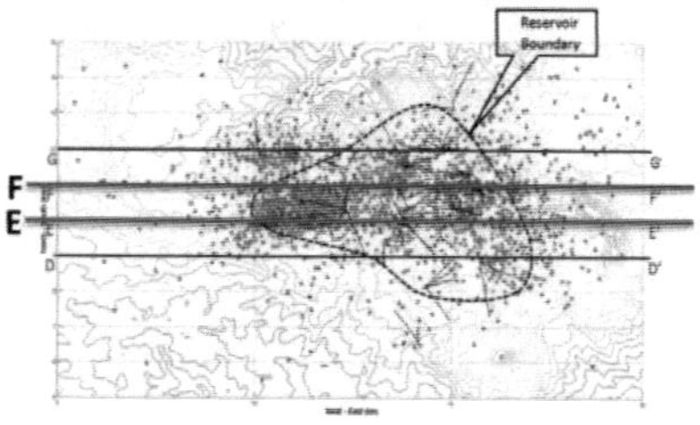

Figure 6. Tomogram of vertical cross-section (North-South) for Vp/Vs ratio anomaly in: 2007, 2008, and 2009 at N-S 10 and 11 km, and well-trajectory (production: black lines, injection: blue lines). Only areas with good resolution are shown. Positive anomalies are shown by blue colour and negative anomalies are shown by red colour.

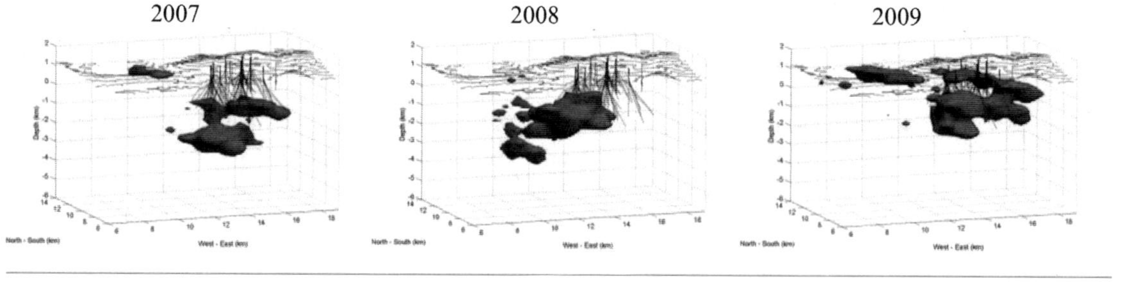

Figure 7. Volumetric of low Vp/Vs anomaly around reservoir area in 2007, 2008, and 2009.

4. Conclusions

Intensive geothermal exploitation at the considered geothermal field in West Java, Indonesia is causing change in local seismic structure which is detectable using four-dimensional seismic tomography in 3-years of observation. The inversion results indicate the presence of low Vp/Vs at around 1.6 in the reservoir area at elevations of -1 to -3 km, interpreted as steam-saturated rock in the reservoir area of this geothermal field. We detect a significant increment of low Vp/Vs anomaly in the strength and volume body from 2007 until 2009. This progressive growth in Vp/Vs ultimately results from depletion of liquid water in the reservoir.

Continuous monitoring of the three-dimensional Vp, Vs, and Vp/Vs ratio is an effective method of reservoir characterization and depletion monitoring at geothermal field, especially at seismically active exploited geothermal area. Four-dimensional or time lapse seismic tomography can potentially provide information about depletion or changes of phase in parts of the reservoir that have not been drilled.

References

[1] Foulger G R 1982 *Geothermics* **11** 259-268
[2] Foulger G R, Grant C C, Ross A and Julian B R 1997 *Geophys. Res. Lett.* **24** 135-137
[3] Ry R V and Nugraha A D 2015 Improve earthquake hypocenter using adaptive simulated annealing inversion in regional tectonic, volcano tectonic, and geothermal observation *AIP Conference Proceedings* **1658** 030004
[4] Stimac J, Nordquist G, Suminar A and Sirad-Azwar L 2008 *Geothermics* **37** 300-331
[5] Start M A 1990 *Trans. Geotherm. Resour. Counc.* **17** 1697-1704
[6] Romero A E, McEvilly T V, Majer E L and Vasco D 1995 *Geothermics* **24** 471-487
[7] Foulger G R, Miller A D, Julian B R and Evans J R 1995 *Geophys. Res. Lett.* **22** 1309-1312
[8] Ry R V and Nugraha A D 2015 High Resolution Delay Time and Shear Wave Splitting Tomography for Reservoir Monitoring of The "RR" Geothermal Field, West Java, Indonesia Using Micro-earthquake Data *4th International ITB Geothermal Workshop*
[9] Wang Z, Batze M L and Nur A M 1990 *Can. J. Explor. Geophys.* **26** 104-112
[10] Takei Y 2002 *J. Geophys. Res.* **107** B2 2043
[11] Gunasekara R C, Foulger G R and Julian B R 2003 *J. Geophys. Res.* **108**(B3) 2134
[12] Kissling E, Ellsworth W L, Eberhart-Phillips D and Kradolfer U 1994 *J. Geophys. Res.* **99** 19635-19646
[13] Um J and Thurber C 1987 *Bulletin of the Seismological Society of America* **77** 972-986
[14] Syahputra A, Nugraha A D and Fatkhan 2012 Development of 3D Seismic Tomography Software: Application to Synthetic Data for Geothermal Exploration *1st ITB Geothermal Workshop*
[15] Fatkhan, Nugraha A D and Syahputra A 2013 *J. Math. Fund. Sci.* **45** 29-38
[16] Widiyantoro S, Gorbatoc A, Kennett B L N and Fukao Y 2000 *Geophys. J. Int.* **141** 747-758
[17] Paige C C and Saunders M A 1982 *ACM Trans. Math. Soft.* **8**(43-71) 195-209

Mapping and Assessing Surface Morphology of Holocene Lava Field in Krafla (NE Iceland) Using Hyperspectral Remote Sensing

M Aufaristama[1,2], Á Höskuldsson[1], I Jónsdóttir[1] and R Ólafsdóttir[1]

[1]Institute of Earth Sciences, University of Iceland, Sturlugata 7, Reykjavík 101, Iceland
[2]Faculty of Geo-Information Science and Earth Observation (ITC), University of Twente, 7500 AE, Enschede, The Netherlands

Email : mua2@hi.is

Abstract. Iceland is well known for its volcanic activity due to its location on the spreading Mid Atlantic Ridge and one of the earth's hot spot. In the past 1000 years there were about 200 eruptions occurring in Iceland, meaning volcanic eruptions occurred every four to five years, on average. Iceland currently has 30 active volcano systems, distributed evenly throughout the so-called Neovolcanic Zone. One of these volcanic systems is the Krafla central volcano, which is located in the northern Iceland at latitude 65°42'53" N and longitude 16°43'40" W. Krafla has produced two volcanic events in historic times: 1724-1729 (Myvatn Fires) and 1975-1984 (Krafla Fires). The Krafla Fires began in December 1975 and lasted until September 1984. This event covered about 36-km^2 surrounding area with lava, having a total volume of 0.25-0.3 km^3. Previous studies of lava surface morphology at Krafla focused on an open channel area by remote sensing are essential as a complementary tool to the previous investigations and to extend the area of mapping. Using Spectral Angle Mapper (SAM) classification approach by selecting spectral reflectance end members, this study has successfully produced a detailed map of the surface morphology in Krafla lava field EO-1 Hyperion (Hyperspectral) satellite images. The overall accuracy of lava morphology map is 61.33% (EO-1 Hyperion). These results show that hyperspectral remote sensing is an acceptable alternative to field mapping and assessing the lava surface morphology in the Krafla lava field. In order to get validation of the satellite image's spectral reflectance, in-situ measurements of the lava field's spectral reflectance using ASD FieldSpec3 is essential.

1. Introduction

The Icelandic landmass was produced by repeated volcanic activities of various type, and nearly all types of volcanoes and styles of eruptions known on earth can be found [1]. Eruption styles range from explosive (produce over 95% of tephra) to effusive (produce over 95% of lava) [2]. Iceland has a high concentration of active volcanoes due to its location on the mid-Atlantic Ridge (divergent tectonic plate boundary) in combination with its position on a volcanic mantle plume located underneath the island. Since the Norse settlement of Iceland in AD 874, 13 of 30 active volcanos in Iceland have erupted [3]. The occurrence of a volcanic eruption every fourth year on average makes Iceland one of the liveliest places in the world with regards to volcanic eruptions. Over the past 500 years, Iceland's volcanoes have erupted a third of the total global lava output [4].

IOP Conf. Series: Earth and Environmental Science **29** (2016) 012002 doi:10.1088/1755-1315/29/1/012002

2. Lava Morphology

Lava morphology is related to the characteristics of the surface morphology of a lava flow after solidification [5]. Morphology of lava is the primary basis for classification of lava flows when rheological properties cannot be directly observed during emplacement [6]. Solidified lava flows and most basaltic lavas can be identified by the terms grouped as pahoehoe, aa, and blocky [6,7]. Aa and pahoehoe lava flows can be found on all of the volcanoes of Iceland [3]. The pahoehoe and aa are by far the most common lava types in Iceland, representing 83% of the 190 lava flows analyzed so far, and fissure-fed pahoehoe are as common as shield forming pahoehoe [3]. In general, on the younger volcanoes, the pahoehoe percentage is higher [8]. Pahoehoe lava has a fairly smooth surface and continuous. Contrary, aa lava is very rough and fragmented. The detail of lava morphology is described in table 1.

Table 1. Description of morphological lava types.

Morphological lava type	Descriptions
Pahoehoe sheet	Smooth surface is divided into sub-horizontal crustal plates, sometimes slightly buckled against each other [9]
Shelly pahoehoe	Shelly pahoehoe is a very vesicular pahoehoe lava type with fragile lava crust [5]. It forms flow lobes and small lava tubes which become hollow inside as lava drains downslope or as the molten lava in the lobe-interior loses gas.
Slabby pahoehoe	This lava is characterized by a flow top of crustal slabs and a pahoehoe base [8]. The slabs are up to several meters across and a few centimetres to decimetres thick [5] [8].
Spinny Pahoehoe	Spiny pahoehoe flows are covered by a rough spiny surface different from the smooth shiny surface of normal pahoehoe [8][9]. Spiny pahoehoe flows commonly form as the last oozes outs of dying pahoehoe flows or stagnating lobes of pahoehoe flows [8]. Spiny pahoehoe also leaks from the edges and the fronts of some aa flows [8].
Cauliflower aa	This is an initial aa lava type in the transformation from pahoehoe to rubbly aa. Protrusions are initially attached to the massive lava beneath but commonly break and form loose debris on the flow surface [9]. Cauliflower aa is common in the shelly and slabby pahoehoe dominated zones where lava flows spilled out after the formation of these morphology [5].
Rubbly aa	Rubbly aa is characterized with a clinkery and blocky surface [5][9]. Surface breccia varies from sand size to blocks several meters in diameter). This lava type has high thermal maturity; the crust during flow is broken by brittle failure [9].

3. Study Area

Krafla is located in northern Iceland, approximately at latitude 65°42'53" N, and longitude 16°43'40" W. The Krafla central volcano has around 8 km wide caldera that was formed in an eruption about 100,000 years ago [11]. Since then, the caldera has widened about 2 km in the East-West direction due to the plate spreading [11]. The Krafla volcano is primarily basaltic, but silicic deposits are found in the vicinity of the caldera [11,12]. Krafla's last eruptive episode, i.e. the "Krafla Fires", resulted 21 tectonic events, and 9 volcanic eruptions [11]. The total area covered with lavas is 36 km^2 and its volume is about 0.25-0.3 km^3 [11]. Morphology of the open channel lava flows in Krafla have been mapped previously [9]. Five flow facies are recognized (1) the initial pahoehoe sheet; (2) slabby pahoehoe and aa; (3) shelly pahoehoe from the channel; (4) rubbly aa; and (5) cauliflower aa (Figure 2a). The previous study was primarily field mapping, video recording and measured pre-flow topography from aerial photographs.

4. Data and Methods

To assess the potential of using hyperspectral remote sensing for identifying lava surface morphology in the Krafla lava fields, the open access satellite data of EO-1 Hyperion was used. The EO-1 Hyperion image used is dated from the 17th of July 2012, with hyperspectral band (242 bands) and 30 meter resolution. The image was geocoded using Ground Control Points (GCP) taken from the aerial photograph and georeferenced into the Lambert Conformal Conic coordinate system on the ISN 1993 Lambert 1993 datum.

The first step of the Hyperion data pre-processing consisted of the elimination of uncalibrated bands. The level 1 Radiometric product has a total of 242 bands, but only 198 of these are calibrated [13]. In this image there are only 170 bands selected due to the uncalibrated bands, bad bands and vertical stripping bands.

Surface reflectance of Hyperion was retrieved using the Fast Line-of-sight Atmospheric Analysis of Spectral Hypercubes (FLAASH) algorithm which is developed in FLAASH ENVI module [13]. The parameterization of ENVI FLAASH includes entering information about the type of sensor and the date of scene, selecting an atmosphere and aerosol model for the correction, and setting the options for the atmosphere correction model. Sub-Arctic Summer Atmospheric model was used in order to characterize the water vapor present in the atmosphere. The aerosol type selected was the rural one in accordance with the Krafla scenario, not strongly affected by urban or any other industrial sources.

The next step, spectra were picked from the image at selected areas that were known to the interpreter from fieldwork and previous studies. Eight spectral endmember points from each morphological type were collected. These include five types of lava morphology pahoehoe, rubbly aa, old lava, cauliflower aa, and shelly pahoehoe. The other classes are, Holocene lava formation, upper Pleistocene formation, and vegetation. Moving average technique was used in order to smoothen the spectral reflectance curves.

Due to the fact that most of the lava in Krafla is basaltic [9], the lava spectral was further validated from a laboratory measurement, obtained from the USGS and provided in ENVI software. This basalt spectral reflectance have five major absorptions, i.e. $Fe3^{+}$ electronic absorptions in oxides/hydroxides, Fe^{2+} electronic absorptions in pyroxene, O–H vibrations in hydroxyl, H_2O and M–OH vibrations in clay minerals (Figure 1) [15]. This USGS sample of weathered basalt has a higher reflectance and also has relatively strong absorptions in the five bands compared to fresh basalt. These spectral will be classified in the next step using Spectral Angle Mapper (SAM).

The Spectral Angle Mapper (SAM) is a classification method which calculates the spectral similarity between the image reflectance spectrums to reference reflectance spectra [16]. In this method the spectral reference is extracted directly from the image endmember selection. This method is not affected by solar illumination factors, because the angle between the two vectors is independent of the vectors length. It takes the arc cosine of the dot product between the test spectrums "t" to a reference spectrum "r"with the following equation [13]:

$$\alpha = \cos^{-1}\left(\frac{\sum_{i=1}^{nb} t_i r_i}{\left(\sum_{i=1}^{nb} t_i^2\right)^{\frac{1}{2}}\left(\sum_{i=1}^{nb} r_i^2\right)^{\frac{1}{2}}}\right) \quad (1)$$

Finally the classification result will be assessed using accuracy assessment. The accuracy assessment shows the percentage difference between classification result and references data. Selections of 150 reference points were based on random equal-stratified sampling.

5. Results and Discussions

According to spectral reflectance curve, the results show that the absorption features of the old lava in Krafla are relatively similar to weathered basalt, except in (Fe/Mg)-OH absorption ranges due to the limited range of remote sensing data and the shifting of absorptions features (Figure 1). Shifting of absorption features in the hyperspectral data could be due to the effect of data smoothing. Figure 1 shows the effect of bad data (850 nm – 1000 nm and 1820 nm – 1951 nm), which causes a flat spectral after data smoothing. The result from the SAM method on EO-1 Hyperion data successfully discriminates the lava morphologies in the Krafla lava field (Figure 2b). The classification based on endmember spectra has an overall accuracy of 61.33% (Table 2). However, the bad data and vertical stripping of the Hyperion images does affect the ability to extract spectral reflectance and identify individual lava morphologies.

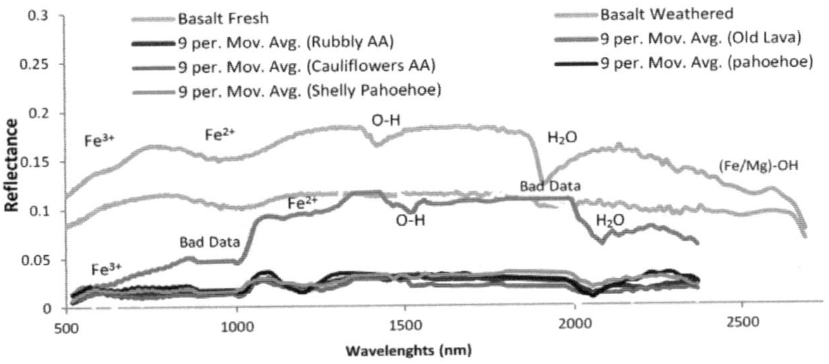

Figure 1. Spectral reflectance of lava morphology comparison between basalt from USGS laboratory measurement and basalt lava in Krafla lava field from remote hyperspectral remote sensing with the absorptions features.

Furthermore, detailed analysis of the individual lava morphology accuracy of EO-1 Hyperion reveals that hyperspectral is prospective to detect the individual lava morphology. In order to be considered, the accuracy needs to be over 50%. Cauliflower aa producer accuracy clearly show that only 43.33% of pixels classified have been identified. According to that result we also need to consider the user accuracy. The user accuracy result of cauliflower aa show 76.47%. Therefore, despite the map produces only 43.33%, a user of this map will find that the map show 76.47% the actual cauliflowers aa. This acceptable user accuracy reveals that the map is still reliable to identify cauliflower aa and the rest of lava morphology.

Table 2. Results of overall, producer and user accuracy including the Kappa statistics for EO-1 Hyperion.

Satellites Image	EO-1 Hyperion	
Lava Classes	User Accuracy	Producer Accuracy
Rubbly aa	68.57%	80%
Cauliflowers aa	76.47%	43.33%
Shelly Pahoehoe	64.29%	30%
Pahoehoe	58.54%	80%
Old Lava	75.86%	73.33%
Overall Accuracy	61.33%	

EO-1 Hyperion produces 30% producer accuracy for shelly pahoehoe. This is the only value for pahoehoe lava morphology for which accuracy falls below 40%. This is due to horizontal strip that occurred in the spectral endmember of shelly pahoehoe and also the vertical strips that appear in the western, centre (main concentration of shelly pahoehoe) and northern part in the Hyperion image. Several other issues may affect the accuracy of lava surface morphology. These include the following: (1) Medium spatial resolution (30 meter) of satellites image; (2) image acquisition differences between references and satellites and (3) small differences of spectral characteristics within lava morphology.

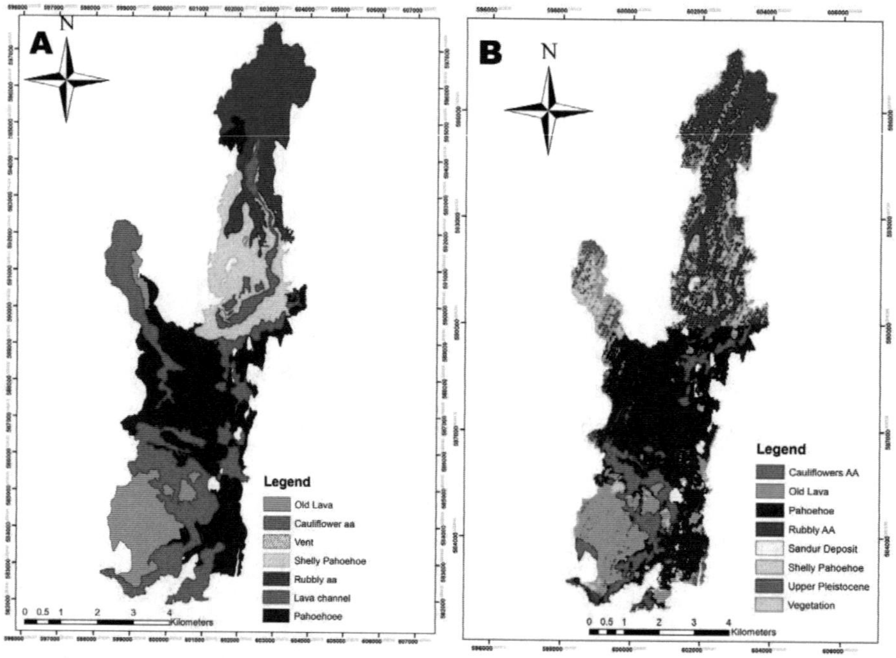

Figure 2. (a) Surface morphology reference of Krafla lava field. **(b)** EO-1 Hyperion SAM classification result successfully produces the lava morphology map within Krafla lava field.

6. Conclusion and Recommendations

Finally, the possibilities of EO-1 Hyperion to identify individual lava flow morphologies using endmember spectra collection, and spectral angle mapper classification have been handled. According to the results, in general hyperspectral remote sensing is capable of mapping and assessing of detail lava morphology in Krafla lava field. The real ground truth in situ measurement of lava field reflectance spectra using ground remote sensing such as ASD FieldSpec3 [17] would be essential to collect from the study area in order to validate the Hyperion images. Spectroscopy laboratory measurements of the Krafla lava field would give the reference of absorptions features of basaltic lava in Krafla lava field. Another option of using airborne hyperspectral image data such as HyMap and AVIRIS could possible to improve detail of reflectance spectra, spatial resolution and quality of data.

Acknowledgements

I am grateful to the LPDP scholarship (Indonesia Endowment Fund for Education) for this research project.

References

[1] Dugmore A, Vésteinsson O 2012 *Black Sun, High Flame, and Flood: Volcanic Hazards in Iceland. Surviving Sudden Environmental Change* pp 67-90

[2] Thordarson T, Larsen G 2007 Volcanism in Iceland in historical time: Volcano types, eruption styles and eruptive history. *J Geodyn.* **43**(1):118–52
[3] Thordarson T, Höskuldsson Á 2008 Postglacial volcanism in Iceland. *Jokull.* (**58**):197–228.
[4] Kozák J, Vladimír Č. Volcanism in Iceland. The Illustrated History of Natural Disasters. Springer Netherlands; 2010. pp 79–83
[5] Murcia H, Németh K, Moufti MR, Lindsay JM, El-Masry N, Cronin SJ, *et al* 2014 Late Holocene lava flow morphotypes of northern Harrat Rahat, Kingdom of Saudi Arabia: Implications for the description of continental lava fields. *J Asian Earth Sci.* **84**:131–45
[6] Kilburn CR 2000 *Lava Flows And Flow Fields. Encyclopedia of Volcanoes.* Academic Press; pp 291–306
[7] Wohletz K, Heiken G 1992 *Volcanology and Geothermal Energy* Berkeley: University of California Press; p 308
[8] Hon K, Johnson J, Gansecki C 2008 *Field Interpretation of Active Volcano : A Handbook for viewing lava.* Reveira T, editor. Hawaii: U.S. Geological Survey Hawaiian Volcano Observatory.
[9] Rossi MJ 1997 Morphology of the 1984 open-channel lava flow at Krafla volcano, northern Iceland. *Geomorphology.* **20**(1-2):95–112
[10] Pedersen GB., Höskuldsson Á, Riishuus M., Jónsdóttir I, Gudmundsson M., Sigmundsson F, *et al* 2015 Nornahraun Lava Morphology and Emplacement: A New Terrestrial Analogue For Planetary Lava Flows. *46th Lunar and Planetary Science Conference.* Houston
[11] Einarsson P 1991 The Krafla rifting episode 1975–1989. Gardarsson A, Einarsson A, editors. Náttúra Mývatns (The Nature of Lake Mývatns). *Reykjavík: Icelandic Nat. Sci. Soc.* 97-139 p
[12] Hjartardóttir ÁR, Einarsson P, Bramham E, Wright TJ 2012 The Krafla fissure swarm, Iceland, and its formation by rifting events. Bull Volcanol.**74**:2139–53
[13] Amici S, Piscini A, Neri M 2014 Reflectance Spectra Measurements of Mt . Etna : A Comparison with Multispectral / Hyperspectral Satellite. Adv Remote Sens. **3**:235–45
[14] Thordarson T, Höskuldsson Á 2014 *Iceland. Classic Geology in Europe 3* 2nd ed. London.: Dunedin Academic Press Ltd pp 171-194
[15] Michalski JR, Kraft MD, Sharp TG, Christensen PR 2006 Effects of chemical weathering on infrared spectra of Columbia River Basalt and spectral interpretations of martian alteration. *Earth Planet Sci Lett.* **248**(3-4):822–9
[16] Girouard G, Bannari A 2004 Validated spectral angle mapper algorithm for geological mapping: comparative study between QuickBird and Landsat-TM. *XXth ISPRS Congress.* pp 599–604.
[17] Robertson KM, Milliken RE, Ruff S, Farmer J, Shock E 2013 Can Vis-Nir Reflectance Spectra Be Used To Assess Formation Environments of Opaline Silica on Mars? *44th Lunar and Planetary Science Conference.* Texas. pp 2–3

IOP Conf. Series: Earth and Environmental Science **29** (2016) 012003 doi:10.1088/1755-1315/29/1/012003

Prediction of landslide run-out distance based on slope stability analysis and center of mass approach

Firmansyah[1]**, S Feranie**[1]**, Adrin Tohari**[2]**, F.D.E Latief**[3]**,**

[1]Laboratory of Earth and Space Science, Physics Department FPMIPA UPI, Jl. Dr. Setiabudhi No. 229, Bandung
[2]Research Center for Geotechnology (P2G-LIPI), Komplek LIPI Jl. Sangkuriang, Bandung
[3]Physics of Earth and Complex System, FMIPA ITB, Jl. Ganesha No. 10, Bandung

E-mail: firmansyah72@student.upi.edu

Abstract. Mitigation of landslide hazard requires the knowledge of landslide run-out distance. This paper presents the application of slope stability analysis and center of mass approach to predict the run-out distance of a rotational landslide model with different soil types. The Morgenstern-Price method was used to estimate the potential sliding zone and volume of landslide material. The center of mass approach used a simple Coulomb friction model to determine the run-out distance. Results of the slope stability analysis showed that the soil unit weight can influence the depth of sliding zone, and the volume of unstable material. The slope model of silty sand and gravel would have the largest volume of unstable mass. From the Coulomb friction analysis, this slope model has higher run-out distance and velocity than other slope models. Thus, the run-out distance will be influenced by soil type and the dimension of unstable soil mass.

1. Introduction

Knowledge about physical properties, mechanism and behavior of landslide is essential for hazard mitigation and risk assessment. Analyses of run-out and slope stability are some important parameter in landslide risk assessments. Landslide-related case studies, theories, and technology advancement have been reported in journal articles, conference proceedings as well as books. One of the analytical theories related to the landslide is the slope stability which based on relationship between driving force and resisting force. The driving force acts on the landslide material in downslope movement which depends on gravity, slope angle, slope material, and water. On the other hand, the resisting force acts oppositely of driving force which depends on shear strength of the slope material. The ratio between the resisting forces to the driving forces is one way to define the safety factor in a force equilibrium analysis [1]. Prediction of landslide run-out distance has also been one of the significant topics in the research field. Currently there are several ways to analyze run-out distance using dynamic models; one of them is center of mass (COM) approach. Prediction of run-out distance using dynamic models has been applied in 2D [2, 3] and in 3D [4, 5] and also statistical analysis [6]. Some researchers [6, 7, and 8] have mentioned that the correlation between volume of landslide material and travel distance is linear. There are several commonly used methods to analyze run-out distance, e.g., the COM method, the end of deposit method [2] and cone fall method [5]. Hungr [4] and Devoli [6] analyzed the run-out distance from initial height higher than 10 m. The aim of this paper is to estimates the run-out distance of rotational landslide in the slope model with initial height less than 10 m. In this model, two soil layers of different type are used to find the influence of soil type to the run-out's parameters (apparent friction angle, travel distance, friction coefficient).

2. Methods

2.1. The Morgenstern-Price Method

There are several methods to determine slope stability. One of the most commonly used methods is the Morgenstern-Price Method (MPM) which firstly proposed by Morgenstern and Price [10]. This method is based on the summation of tangential and normal forces to the base of a slice and the summation of moments about the center of the base of each slice. The relationships for the base normal force (*N*) and inter-slice forces (E, T) are shown in figure 1.

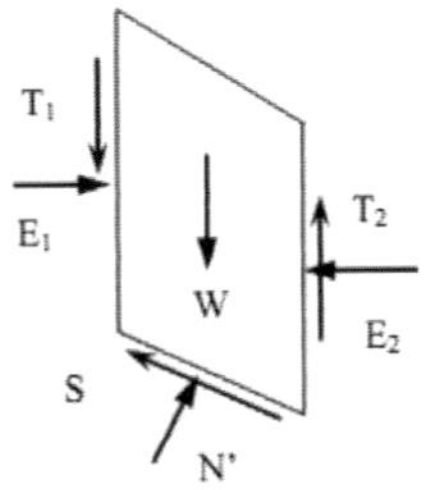

Figure 1 Force diagram of inter-slice forces in Morgenstern-Price method [1]

where *W* is the weight of a slice, *S* is the shear mobility force (friction) on the base of each slice, *T* is the vertical inter-slice shear forces, *E* is the horizontal inter-slice normal forces, and *N* is the total normal force on the base of a slice. The expression of safety factor (*F*) in terms of force (F_f) and moment equilibrium (F_m) are given as follows:

$$F_f = \frac{\sum\left[\left\{c'l + \left(N - ul\right)\tan\phi\right\}\sec\alpha\right]}{\sum\left\{W - \left(T_2 - T_1\right)\right\}\tan\alpha + \sum\left(E_2 - E_1\right)} \tag{1}$$

$$F_m = \frac{\sum\left(c'l + \left(N - ul\right)\tan\phi\right)}{\sum W\sin\alpha} \tag{2}$$

where c', ϕ, α, l is cohesion, friction angle, inclination of slip surface, and slice base length respectively in effective stress terms of each slice base.

2.2. Center of mass (COM) approach to predict run-out distance

In this study, we combine the COM approach with the end of deposit method to predict the run-out distance of landslide material which also characterized by the slope stability analysis. Landslide movement can be determined using a simple geometric rule known as energy budget calculation from starting point based on a simple Coulomb frictional model as implemented in this study. This principle was modified and applied without volume dependency. The detail about this approach can be seen in figure 2a. The variables used to calculate velocity and run-out distance are H_G (the altitude of COM of the starting zone), $h(x)$ (the altitude of the topography at the distance x from the start zone), m (the mass), μ (the friction coefficient), ϕ (the internal friction angle), and L_G (the deposit distance from starting zone).

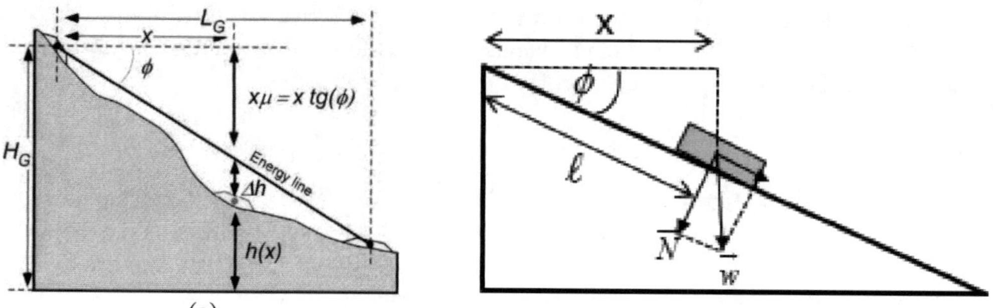

(a) (b)

Figure 2. (a) Variables used to calculate velocity and energies based on the energy line concept [11]. (b) A simple Coulomb friction model

In simple Coulomb friction model (figure 2b), the friction is proportional to the normal force N which also related to the weight w by:

$$\left|\vec{N}\right| = \left|\vec{w}\right|.\cos\left(\phi\right) \tag{3}$$

and the energy loss by friction is given by,

$$\mu = \tan\left(\phi\right) \quad \cos\left(\phi\right) = \frac{x}{l} \quad \rightarrow \; l = \frac{x}{\cos\left(\phi\right)}$$

$$E_f = \mu\left|N\right|.l = \mu.m.g\cos\left(\phi\right).\frac{x}{\cos\left(\phi\right)} = m.g.x.\tan\left(\phi\right) \tag{4}$$

Hence, the balance of energy at point x of landslide from an elevation H is given by [2]:

$$mg\left(H - h\left(x\right)\right) = \tfrac{1}{2}mv_x^{\,2} + mgx\tan\left(\phi\right) \tag{5}$$

Velocity expression for any x-position can obtained from equation (5):

$$\tfrac{1}{2}mv_x^{\,2} = mg\left(H - h\left(x\right) - x\tan\left(\phi\right)\right) \rightarrow Let: \Delta h\left(x\right) = \left(H - h\left(x\right) - x\tan\left(\phi\right)\right)$$

$$v(x) = \sqrt{2g\Delta h\left(x\right)} \tag{6}$$

where g is gravity acceleration and $v(x)$ is the velocity at point x.

Rearranging equation (5) to estimate the stopping point horizontal distance (run-out distance) x_{stop} by selecting $v(x) = 0$, and using $\mu = \tan\left(\phi\right)$, we obtain the expression of run-out:

$$L = x_{stop} = \frac{H_G - h\left(x_{stop}\right)}{\tan\left(\phi\right)} \tag{7}$$

In general form, the equation (7) can be simplified as,

$$L = \frac{H}{\tan\left(\phi\right)} \tag{8}$$

2.3. *Soil Layers Model*

Safeland [12] generally classified run-out model for landslide hazard and risk mapping into two main models, i.e., empirical models and rational models. In rational model there are sub-models such as a continuum models (3D models based on mixture theory, velocity-pressure models, depth integration model) and a discrete models, where in this study we used the discrete models. The soil layers model was constructed based on the assumptions that there are two different layers. The first one is the upper layer which is an unstable zone that acts as sliding mass and the second one is the lower layer which is the stable zone and more consolidate than upper layer. In this study, it is assumed that each layer is composed by homogeneous soil type, and the slip surface is assumed circular which is suitable to be used in the MPM. The difference of soil type is indicated from unit weight/bulk weight (kN/m^3). The elevation model is based on an unpublished work which varies between 576 m a.s.l at the toe and 610 m a.s.l at the top (figure 3a). After the model was constructed, the parameters are then used (bulk weight, cohesion, angle of shearing) in measuring the safety factor using MPM to obtain the most critical safety factor. In figure 3b, green indicates the slide mass and red indicates the safe zone.

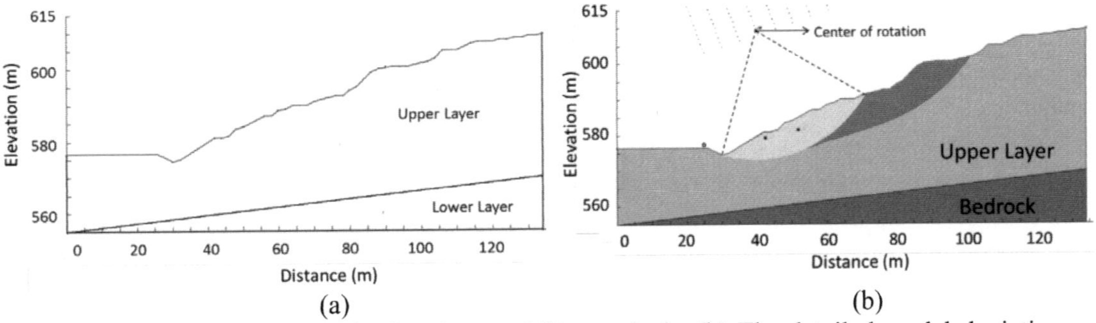

Figure 3. (a) Slope model used in the slope stability analysis. (b) The detailed model depicting the center of rotation. The upper layer is composed by homogenous soil type.

3. Results and Discussion

3.1 Safety factor analysis

The unit weight of soil type used in this study is based on an existing geotechnical data and the parameters to determine the safety factor such as angle of shearing is assumed constant and the variation of the cohesion is used to obtain the most critical factor (see table 1). Using model in figure 3 (b), the prediction of run-out distance as well as the velocity can then be carried out.

Table 1. Parameters used in the safety factor analysis

Parameters	Units	Soil type (upper layer)		
		(a) silty sand	(b) sand, dense, uniform	(c) silty sand and gravel
Total Volume	[m^3]	267.290	267.290	308.150
Total Resisting Moment	[kNm]	48.887	57.856	1.19×10^5
Total Activating Moment	[kNm]	45.020	55.765	1.14×10^5
Total Resisting Force	[kN]	1196.900	1418.200	2350.200
Total Activating Force	[kN]	1102.400	1367.100	2240.500
Unit Weight γ	[kN/m^3]	13.91	17.23	24.5
Angle of Shearing	[°]	10	10	10
Cohesion	[kPa]	15	17	25
Safety factor	-	1.086	1.037	1.049

Based on model in figure 3 and the corresponding parameters in table 1, the volume of the landslide material of soil type (a) and (b) are equal and smaller than soil type (c). The results show that the unit weight is strongly affects the slope stability: the heavier unit weight is associated with the greater driving force.

3.2 Run-out prediction

Based on figure 3 (b), the center of mass position of the green zone can be estimated. The center of mass is then applied on the simple Coulomb friction model to predict the landslide run-out. The parameters used in the run-out analysis are listed in table 2. Figure 4 visualize the run-out analysis, and the prediction of run-out distance as well as the velocity can also be obtained from these images.

Table 2. Parameters used in the run-out analysis and the calculated results (L and v).

Soil Type of Upper Layer Model	Fall of Height (H)	Friction Angle (ϕ)	Friction Coefficient (μ)	Run-out (L)	Velocity (v)
(a) Silty Sand	4.6 m	10°	0.176	26.09 m	3.08 m/s
(b) Sand, dense, uniform	3.5 m	13°	0.141	24.90 m	2.47 m/s
(c) Silty sand and Gravel	5.4 m	13°	0.141	38.42 m	4.59 m/s

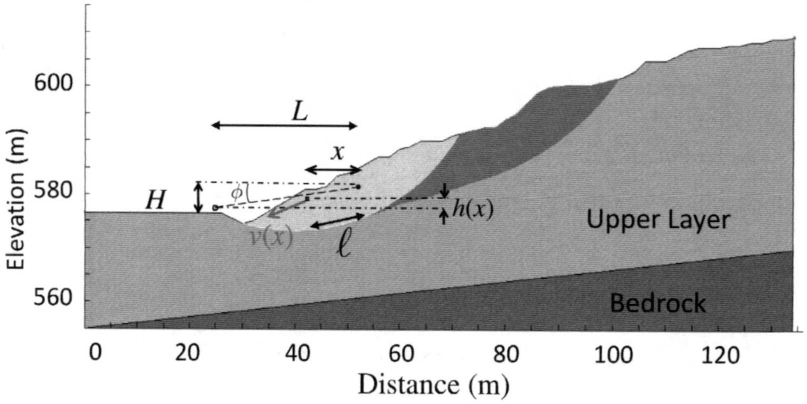

Figure 4. Geometrical parameters used in run-out and velocity analysis.

Based on table 2, soil type (a) and (b), the predicted run-out distance are equal but smaller than type (c). However, these results depend on the determination of center of mass position. From these results, the slide mass of third soil type is larger than the other ones, which can cause greater damage to the surrounding environment. From calculation of the run-out distance, it is predicted that the third soil type can reach further than the other ones about 38.42 m from the initial point of center of mass.

This center of soil mass is predicted to reach velocity of 4.59 m/s, which is faster than other soil types. Soil type of silty sand at point x is predicted to reach velocity about 3.08 m/s and soil type of sand, dense, uniform at point x is predicted to reach velocity about 2.47 m/s. These predictions of the velocity are considered very low compared to the results from several previous researches [3, 4]. This might be caused by initial height of center of mass which is lower than 10 m and other parameters such as apparent friction angle and friction coefficient. However the results indicate that unit weight is influence velocity and run-out distance. The correlation between unit weight and run-out's parameters are shown in figure 6. Based on figure 6a, friction coefficient (μ) will decrease when unit weight (γ) increase, it is consistent with the previous results [6, 7, 8, 9]. From figure 6b, we find that heavier landslide material produce larger value of the run-out distance.

4. Conclusions and future work

From the analyses, we can conclude that soil type and the dimension of unstable soil mass can influence run-out distance (L) of rotational landslide in the soil slope model which the biggest of unit weight and the largest of volume of unstable soil mass in this study (silty sand and gravel) has higher velocity and further run-out distance than the other soil type.

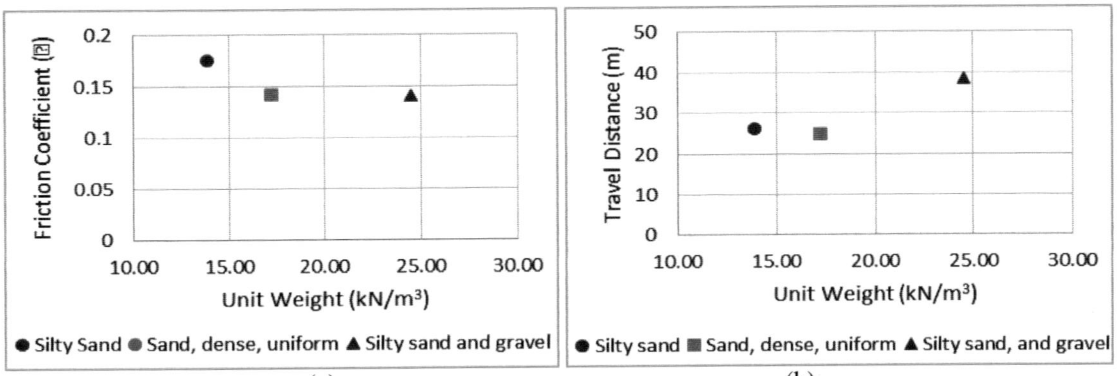

(a) (b)

Figure 5. The correlation between parameters (a) unit weight and friction coefficient; (b) unit weight and travel (run-out) distance.

The analyses above are based on the situation where a failure occurs caused by a certain value of cohesion which produces the most critical safety factor, and the approaches used in this study could not estimate the distribution of landslide material. Future work will use geotechnical analysis (laboratory test) to obtain actual data of physical properties of soils. We also devise a research plan to predict the deposit distribution and to estimate the instantaneous velocity of the landslide movement from the beginning of the slide until it is fully deposited.

Acknowledgement
The research was made possible by funded by the research program "Desentralisasi DIKTI-ITB 2015"

References

[1] Krishna P A 2006 Slope Stability Evaluations by Limit Equilibrium and Finite Element Methods *PhD Thesis at Norwegian Univ. of Science and Technology* Trondheim: April 2006
[2] A Heim 1932 Bergsturz and Menschenleben *Fretz und Wasmuth* Zurich pp 218 (in Germany)
[3] Che-Ming Y, Wei-Lun Y, Chih-Yu K, Toshihiko S, Chyi-Tyi L, Tetsuhiro T, and Yuki M 2014 Initiation, movement, and run-out of the giant Tsaoling landslide-What can we learn from a simple rigid block model and a velocity-displacement dependent friction law? *Engineering Geology* **182** pp 158-181
[4] O Hungr 1995 A model for the run-out analysis of rapid flow slides, debris flows, and avalanches *Can. Geotech. J.* **32** pp 610-623
[5] M Jaboyedoff and V Labiouse 2011 Technical Note: Preliminary estimation of rockfall run-out zones *Nat. Hazards Earth Syst. Sci.* **11** 819-828 doi: 10.5194/nhess-11-819-2011
[6] G Devoli, F V De Blasio, A Elverhoi, and K Hoeg 2009 Statistical Analysis of Landslide Events in Central America their Run-out Distance *Geotech. Geol. Eng*, **27** pp 23-42 O
[7] Hungr 1981 Dynamics of rock avalanches and other types of slope movements *Ph.D. thesis, University of Alberta*, Edmonton
[8] J Corominas 1996 The angle of reach as a mobility index for small and large landslides *Can. Geotech. J.* **33** pp 260-271 doi: 10.1023/B:NHAZ.0000007094.74878.d3
[9] D Rickenmann 1999 Run-out Prediction Method *In M. Jacob and O. Hungr (eds) Debris-flow Hazards and Related Phenomena*
[10] N R Morgenstern and V E Price 1965 The Analysis of the Stability of General Slip Surface *Geotechnique* **Vol 15** (1) pp 77-93
[11] M Jaboyedoff, P Horton, A Loye, and A Pedrazzini 2008 Run-out-empirical approaches *in Workshop Barcelona 2008 Mountain Risks* September 2008
[12] Safeland 2010 Recommendations for run-out models for use in landslide hazard and risk mapping *Deliverable 1.9*

IOP Conf. Series: Earth and Environmental Science **29** (2016) 012004 doi:10.1088/1755-1315/29/1/012004

A comparison of flash flood response at two different watersheds in Grenada, Caribbean Islands

R A Pratomo[1], V Jetten[2], and D Alkema[3]
[1] Urban and Regional Planning, Tadulako University, Palu, Indonesia
[2,3] Earth System Analysis, Faculty of-Geo Information Science and Earth Observation (ITC), University of Twente, Enschede, the Netherlands

E-mail: arispratomo05@gmail.com

Abstract. Grenada is one of the susceptible areas to flooding. It is due to the high intensity of rainfall in each year and Grenada often hit by hurricanes and tropical storms. Flash-flood often occur in Grenada, specifically in two areas (Gouyave and St. John′ s watershed). Both of them have different characteristics. Gouyave watershed represents rural area, whereas St. John′ s watershed is an urban area. This paper aims to understand the flash-flood response of Gouyave and St. John′ s watershed in different return periods. It is emphasized that urbanization is an important factor related to flash-flood. This paper uses quantitative methods with flood modelling using OpenLISEM software. Input data to develop flood modelling are DEM (Digital Elevation Model), saturated hydraulic conductivity (Ksat), initial soil moisture, surface roughness (Manning's n), and random roughness. The result shows that St. John′ s watershed is more sensitive and has higher response to flash-flood than Gouyave. St. John′ s watershed is more urbanized which decreases water infiltration. So, it increases the potential run-off and flash-flood events become more massive.

1. Introduction

Flood is one of the most frequent natural disasters worldwide which usually occur in low land areas, near the rivers, coasts, or some areas which have high susceptibility to flooding. One of these susceptible areas is Grenada, Caribbean Islands. Flash-flood events in Grenada are caused by the local heavy rainfall, such as the latest event in 2011. In 2011, the intensity of rainfall in this area is 2.350 mm/year [1]. The high intensity of rainfall in certain areas was also delivered by tropical storms and hurricanes. Grenada's area is called *the hurricane belt* because it is often hit by hurricanes, like hurricane Janet in 1955, hurricane Flora in 1963, hurricane Ivan in 2004, and hurricane Emily in 2005. The hurricane Ophelia in 2011 was one of the most damaging events because it triggered the flash-flood event.

Flash-flood in Grenada usually occurs in two watershed areas, Gouyave watershed and St. John's watershed. In Gouyave watershed, flash-flood hits the downstream area of Charlotte River, while it hits the downstream area of St. John's River in St. John's watershed. Flash-flood event is usually

[1] To whom any correspondence should be addressed.

triggered by the intensity of rainfall in both of areas. During the rainy season on 2011, the intensity of rainfall in Gouyave watershed is 250-350 mm, while 150-250 mm in St. John's watershed [1].

Both of watersheds have different characteristics. St. John's watershed represents an urban area which is characterized with high intensity of built-up area. St. George, the capital city of Grenada is located in this area and the urban development is growing faster. However, Gouyave is still dominated by natural land-use types, such as forest, mix-trees, bare, and agriculture lands. It seems that Gouyave watershed represents rural area. Flash-flood conditions in these two watersheds are shown at Figure 1 and Figure 2.

Figure 1. Flood condition in Gouyave watershed

Figure 2. Flood condition in St. John's watershed

The different characteristics of those watersheds would influence to the flash-flood response in each watershed. Thus, this paper aims to understand the flash-flood response at Gouyave and St. John's watershed in different return periods which is emphasized that urbanization is an important factor related to flash-flood. The understanding of urbanization factors and their impacts on flash-flood can help to develop the Grenada's disaster risk reduction strategies.

2. Method

This paper uses quantitative methods with flood modelling using OpenLISEM software. OpenLISEM is an event based natural hazard model to simulate runoff, erosion and flash floods for a single rainfall event [2, 5]. It was used to build a flood modelling with generate the land-use change and its impacts to flood events, for example land-use change and its relationship to flood. The data requirements for developing flood modelling were:

- Catchment characteristics, include local drain direction, catchment boundaries, area covered by rain gauges, slope gradient, locations of outlet and sub-outlets (derived from DEM.map). DEM.map was created based on the interpolating contour line from topographic map with resolution 20 meters. Besides that, it was required rainfall data (from Maurice Bishop International Airport, Grenada).
- Vegetation characteristics, such as leaf area index (LAI), fraction of soil covered by vegetation (PER), and vegetation height (CH), were identified from field observation.
- Soil surfaces, include random roughness (RR) and surface roughness (Manning's n) which were adapted from literature.
- Channel data, including local drain direction of channel network, channel gradient, Manning's for the channel, width of channel scalar, and channel cross section shape.
- Green and Ampt Layer is a physical model to simulate infiltration rates to measurable soil properties, such as saturated hydraulic conductivity (Ksat), saturated volumetric soil moisture content (Thethas), initial volumetric soil moisture content (Thethai), and soil depth [8]. There were two layers data. The first layer was derived from field observation and literature and the second layer was obtained from *saxton equations* and field observation. *Saxton equations* is a software for estimating soil water characteristics for water potential and hydraulic conductivity based on soil textures, organic matters, and structures [7].

Flood modelling was derived in two layers to reduce the uncertainty in values of soil properties, such as saturated hydraulic conductivity (Ksat), initial soil moisture, and porosity. The measurement of those soil properties was conducted during field observation in Grenada. The result shows that the top-soils contain much organic materials, specifically roots and leaves, so their values were high. Thus, this paper used two layers; the first layer was obtained from field observation, while the second layers were adapted from literature.

3. Study Area

Grenada is one of the small island states in the Eastern part of the Caribbean region, specifically between the Caribbean Sea and the North Atlantic Ocean. It is located between latitudes 11° 59' and 12° 20' N and longitudes 61° 36' and 61° 48' W. Total area of Grenada's main island is 312 km^2 with total population of 110.000 people [4]. Orientation map of Grenada are shown in **Figure 3**.

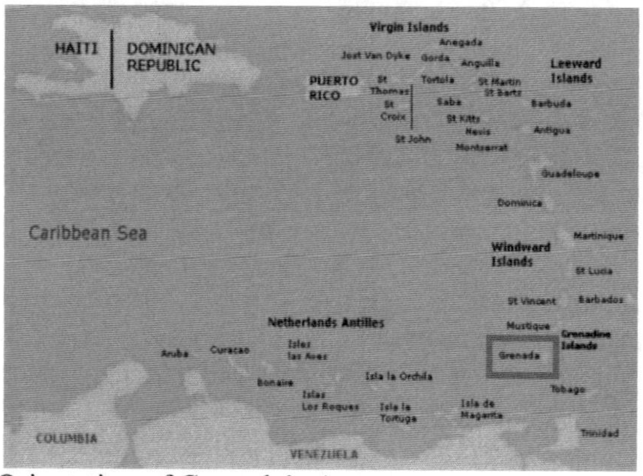

Figure 3. Orientation of Grenada's Area

The main island of Grenada has 71 watersheds and 12 of them are large watersheds. From 12 largest watersheds in Grenada, St. John's watershed and Gouyave watershed are selected as study area (**Figure 4**). Most areas at Gouyave watershed are located in the elevation of more than 200 m above sea level, specifically in the eastern parts and southern parts of Gouyave, while, St. John's watershed is dominated by flat area with elevation between 0-50 m above sea level. Both watersheds are dominated by steep-slope (25-55%).

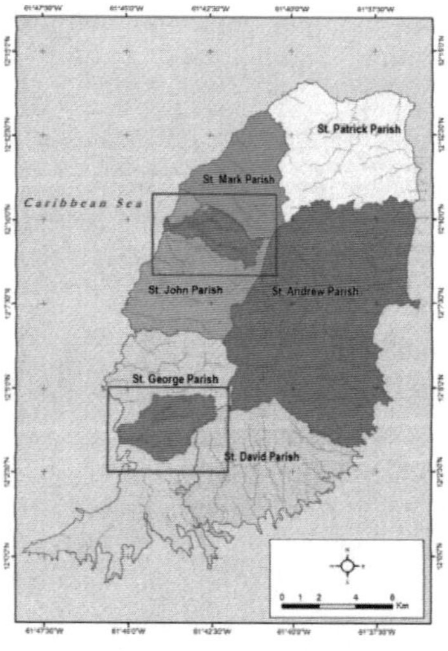

Figure 4. Orientation map of Gouyave and St. John's watershed. Gouyave watershed is located in St. John Parish, while St. John's watershed is located in St. George Parish.

Gouyave watershed is dominated by natural land-use types (**Figure 5**). 56,96% of Gouyave watershed are covered by mixed tree. Besides that, there are also forest, agriculture lands and bare land. The built-up areas can be found in their coastal area (5,53%). Meanwhile, St. John's watershed has higher intensity of built-up areas (**Figure 6**). They cover 23,28% of total area. Moreover, there are many land conversions in this area due to the urban developments and the increasing of urban population.

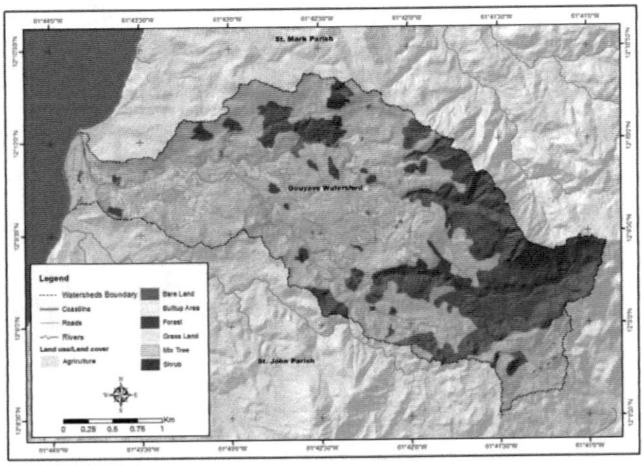

Figure 5. Land-use map of Gouyave watershed

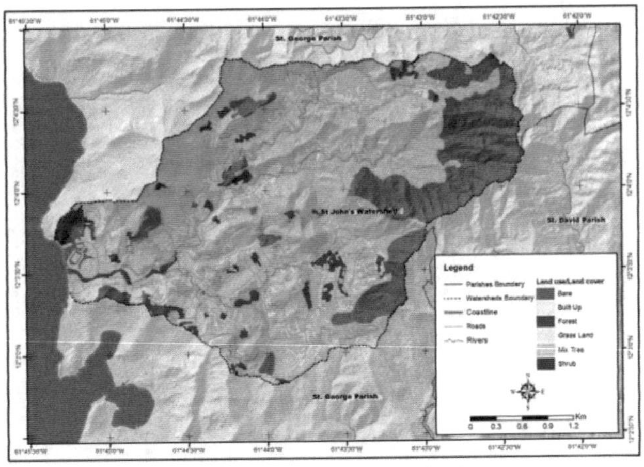

Figure 6. Land-use map of St. John's watershed

4. Method

4.1 Flood Modelling

Flood modelling was conducted in the different return periods. Those return periods were determined from Gumbel Method. Gumbel method is a method to predict the return period of rainfall based on rainfall history data, describe the occurrence probability of extreme rainfalls, and identify the distribution of extreme values [3]. The result of Gumbel Method analysis is shown in **Table 1**.

Table 1 Return periods and maximum daily rainfall

Return Period (years)	Max daily rainfall (mm)	Remarks
2	75.64	Maximum daily rainfall in 2013 (low-categories of rainfall)
5	110.20	It is only used to identify the sensitivity of two watersheds to response flash-flood.
10	133.08	
35	172.42	Maximum daily rainfall in 2011. It is used as an event-based condition and basic data for data calibration and validation

Source: analysis, 2015

Besides the intensity of rainfall, there were some important data that were used for modelling show at **Table 2**.

Table 2 Soil physical parameters and vegetation parameters

Land-use Types	Parameters						
	Ksat (mm/hr)	Porosity	Manning's n (cm)	RR (cm)	PER	CH (m)	LAI
Agriculture	10.714	0.585	0.08	0.2	0.35	4	1.08
Bare land	12.86	0.495	0.023	0.05	0.2	0.05	0.56
Built-up area	10	0.4	0.03	0.05	0.1	3	0.263
Forest	69.643	0.578	0.1	0.5	0.8	13	4.02
Grass land	19.29	0.58	0.023	0.2	0.85	0.5	4.74
Mix tree	42.857	0.554	0.1	0.5	0.7	10	3.01
Shrub	15.96	0.51	0.04	0.2	0.4	1.2	1.28
Road	0	0	0.001	0.05	0	0	0
Water	0	0	0.033	0.01	0	0	0

Source: analysis, 2015

The result of flood modelling is flash-flood depth. The flash-flood depth in both watersheds will be compared to know the sensitivity of watershed response. It is explained in section 5.

4.2 Model Calibration and Validation

Flood model calibration was conducted to minimize the deviation between the result of flood modelling and the observed data. In this paper, calibration was done based on flood depth data which were obtained from Grenada's flash-flood events in 2011.

Flood modelling was calibrated by changing and combining three sensitive parameters, such as Ksat, porosity, and Manning's n [6]. Then, the result of model calibration was evaluated using model Bias and Root Mean Square Error (RMSE). The formula of model Bias and RMSE are:

$$\text{Bias} = \frac{\sum_{i=1}^{n}(y1-x1)}{n} \quad (1)$$

$$\text{RMSE} = \sqrt{\frac{\sum_{i=1}^{n}(y1-x1)^2}{n-1}} \quad (2)$$

Bias was calculated from the differences of mean values between paired observed and simulated values, while RMSE was measured to show the deviation between resulting values and observed values. The result of model calibration shows that the bias value is 0.123 m and RMSE value is 0.303 m. These values are lowest and both of them are close to zero. Based on this calibration, this configuration is used as model validation in St. John's watershed. The result of model validation is 0.170 m (bias value) and 0.260 m (RMSE value). It is implied that flood modelling has a good performance.

5. Result and Discussion

The comparison of flood modelling conducted to understand how sensitive both of watersheds to response flood. The comparison is shown at Table 3.

Table 3 Watershed response to flash-flood in different return periods

Characteristics	Gouyave watershed				St.John's watershed			
	2 years	5 years	10 years	35 years	2 years	5 years	10 years	35 years
Flood depth max (m)	0.00	0.00	0.21	1.74	0.00	2.75	2.95	3.35

Source: analysis, 2015

Table 3 shows that both watersheds are safe from flash-flood events in 2 years return period. However, in the 5 years return period, flash-flood start to occur in St. John's watershed with maximum flood depth of 2.75 m (**Figure 7**). Meanwhile, there does not occur flash-flood in Gouyave.

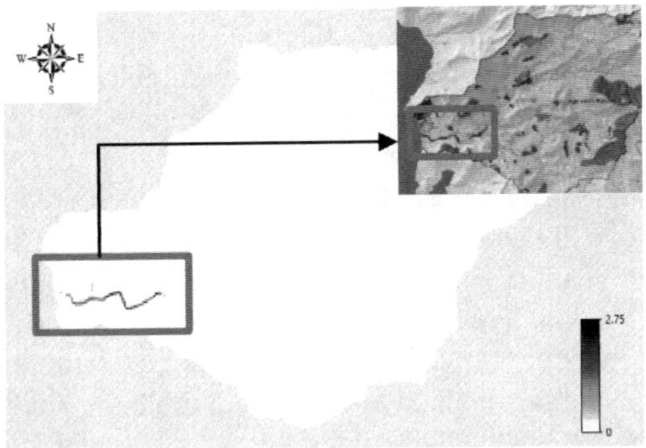

Figure 7. Flood depth of St. John's watershed (m) at 5 years return period

Gouyave watershed starts to respond flash-flood in 10 years return period with 0.21 m of flood depth. It shows that St. John's watershed is more sensitive and has higher response to flash-flood. The flood depth in St. John's watershed is higher than Gouyave (**Figure 8 and 9**).

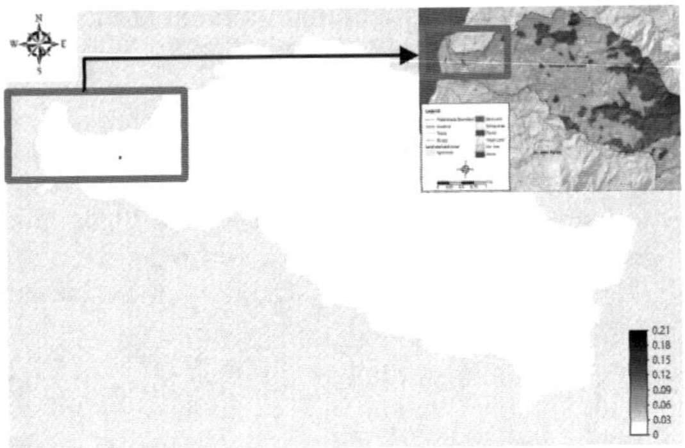

Figure 8. Flood depth of Gouyave watershed (m) at 10 years return period

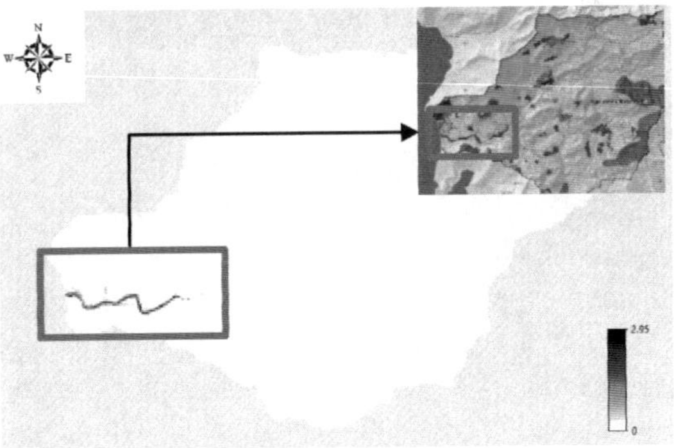

Figure 9. Flood depth of St. John's watershed (m) at 10 years return period

In 35 years return period, flash-flood events in St. John's watershed are more massive than Gouyave. The flood depth of St. John's watershed is also higher (3.35 m) than Gouyave (1.74 m) in the same period (**Figure 10 and 11**).

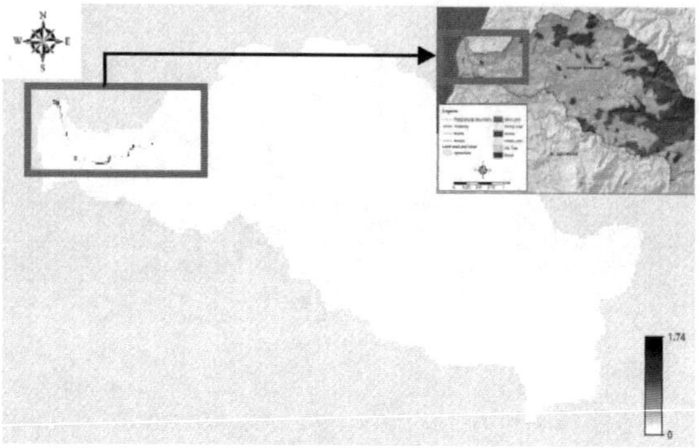

Figure 10. Flood depth of Gouyave watershed (m) at 35 years return period

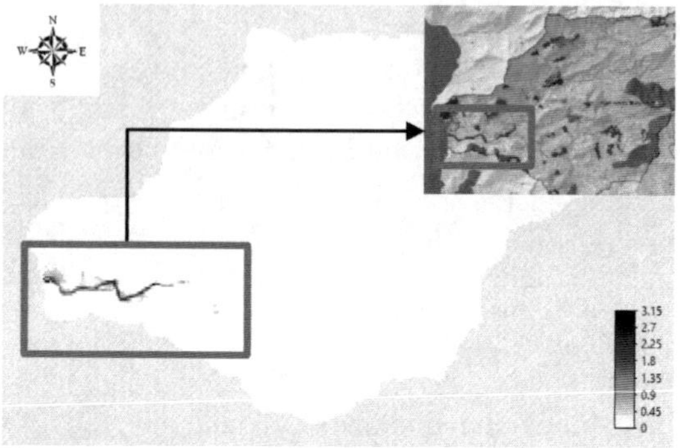

Figure 11. Flood depth of St. John's watershed (m) at 35 years return period

The difference of land-use characteristics affects flash-flood response. St. John's watershed is more urbanized. The urban development in St. George city (capital city of Grenada) increase rapidly, such as increasing the numbers of residential area, commercial area, public facilities, and others. The development of built-up areas in this watershed not only occurs in downstream area, but also in upstream area (**Figure 6**). The increase of built-up areas decreases the capacity of water infiltration into the soil. Thus, run-off will flow down into downstream area and causes flash-flood.

Less water infiltration is influenced by Ksat. The saturated hydraulic conductivity (Ksat) rate of water to infiltrate into the soil. Processes of soil compaction and human interventions with heavy machines in built-up area change the soil pores and make soil more compact, so the Ksat value will decreases. The Ksat value of built-up area is lower than others (10.00 mm/hr). Meanwhile, the natural land-use types, such as forest and mixed-tree have high value of Ksat. Ksat value of forest is 69.64 mm/hr and mixed-tree is 42.85 mm/hr (**Table 2**). If Ksat value decreases, the water infiltration rate will decrease too.

Generally, slope affects the flash-flood response. The greater slope makes the infiltration rate lower. However, slope in both watershed areas are same dominated with 25-55%. It means that slope factor is not a major indicator to determine the flash-flood response.

Besides land-use characteristics, shape and size of watershed also influence run-off [9]. Shape of St. John's watershed is rounder and bigger than Gouyave watershed. Run-off from multiple locations in St. John's watershed flow down and arrive at the same time in downstream areas. It causes the peak flow in St. John's watershed to be is higher than Gouyave watershed and gives the greater impacts to this area.

6. Conclusion

Urbanization is a factor that influences the flash-flood response. The high intensity of built-up areas decreases the water infiltration and increases the potential run-off. It is shown at St. John's watershed. This area responds by flash-flood in 5 years return period, meanwhile Gouyave watershed responds in 10 years return period. The flood depth of St. John's watershed is also higher than Gouyave in the same return period. It means that an area which is more urbanized has a higher response to flash-flood and has greater impact to flash-flood events.

7. References

[1] CEHI 2006 *Geographical Information System-Assisted Water Availability Analysis for Grenada* (St.Lucia: UNEP)

[2] De Roo A P J and Victor Jetten 1999 Calibrating and validating the LISEM model for two datasets from the Netherlands and South Africa *Catena* **37** 477

[3] El-Naqa Ali and Nasser Abu Zeid 1993 A Program of frequency analysis using Gumbel's Method *Ground Water* **31** 1021

[4] GFDRR 2010 *Disaster Risk Management in Latin America and the Caribbean Region: Grenada* (Washington DC)

[5] Jetten V and De Roo APJ 2001 Spatial Analysis of Erosion Conservation Measures with LISEM in Landscape Erosion and Evolution Modelling ed Harmon RS, Doe WW (New York: Kluwer Academic) p 429-445

[6] Prachansri Saowanee 2007 Analysis of soil and land cover parameters for flood hazard assessment: a case study of the Nam Chun watershed, Phetchabun, Thailand (Enschede, the Netherlands: International Institute for Geo-information Science and Earth Observation, Master of Science in Geo-information Science and Earth Observation, specialization: Geo-hazards)

[7] Saxton K E and W J Rawls 2006 Soil water characteristics estimates by texture and organic matter for hydrologic solutions *Soil science society of America Journal* **70** 1569

[8] Setiawan A 2010 Study of Land-Use Change Effect on the Run-off using LISEM Rainfall Model (Enschede, the Netherlands: Faculty of-Geo Information Science and Earth Observation (ITC), University of Twente

[9] US National Oceanic and Atmospheric Administration 2010 *Flash Flood Early Warning System Reference Guide* (United States: National Weather Service International Activities Office)

IOP Conf. Series: Earth and Environmental Science **29** (2016) 012005 doi:10.1088/1755-1315/29/1/012005

Permeability Estimation of Porous Rock by Means of Fluid Flow Simulation and Digital Image Analysis

C W Winardhi[1], F I Maulana, F D E Latief
Department of Physics, Faculty of Mathematics and Natural Sciences,
Institut Teknologi Bandung, Jalan Ganesha 10, Bandung, Indonesia

E-mail: chandra.winardhi@s.itb.ac.id

Abstract. Permeability plays an important role to determine the characteristics of how fluids flow through a porous medium which can be estimated using various methods. Darcy's law and the Kozeny-Carman equation are two of the most utilized methods in estimating permeability. In Darcy's law, permeability can be calculated by applying a pressure gradient between opposing sides of inlet-outlet of a certain direction. The permeability then depends on the fluid viscosity and the flowrate. The Kozeny-Carman equation is an empirical equation which depends on several parameters such as shape factor of the pore, tortuosity, specific surface area, and porosity to determine the permeability. For both methods, digital image obtained by means of Micro CT-Scan is used. In this research, the permeability estimation using the Darcy's law was conducted by simulating fluid flow through the digital image using Lattice Boltzmann Method (LBM). As for the Kozeny-Carman equation, digital image analysis was used to obtain the required parameters. Two Kozeny-Carman equations were used to calculate the permeability of the samples. The first equation (KC1) depends on pore shape factor, porosity, tortuosity, and specific surface area while the second equation (KC2) only depends on pore radius, porosity, and tortuosity. We investigate the methods by first testing on three simple pipe models which vary in the radii. By using the result from Darcy's law as a reference, we compare the results from the Kozeny-Carman equations. From the calculation, KC2 yield smaller difference to the reference. The three methods were then applied to the Fontainebleau sandstone to verify the previous result.

1. Introduction

Permeability is one of petrophysical properties of a sedimentary rock that is significant in evaluating the quality of a reservoir. Permeability quantitatively describes the flow characteristic of fluids which flow through a porous medium. Permeability can be measured using laboratory-based measurement apparatus on core plug samples from well core. Filomena et al. [1] describe the advantages of using gas-driven permeameters, i.e., the measurement can be quickly performed and the samples are not contaminated. In some cases such as in a clay-bearing samples, the gas permeameter does not affect the sample, in contrary to the liquid driven ones which might swell and destroy the sample. Alternatively, permeability can also be calculated indirectly by utilizing a digital image of a porous rock sample. Permeability can then be calculated on the sample by means of simulating fluid flow through the sample using methods such as Lattice Boltzmann Method (LBM), or it can also be

[1] To whom any correspondence should be addressed.

estimated using various empirical equations by first evaluating the physical parameters of the digitized data of the sample using Digital Image Analysis (DIA). LBM uses Darcy's principle in calculating permeability, while DIA can be applied to numerous empirical equations such as the well-known Kozeny-Carman (KC) equations. In this study, we compare the result of permeability calculation using LBM and the permeability estimation using two empirical equations derived from the KC equations. Performance and accuracy of the methods were evaluated to analyze the applicability of the methods on various cases.

2. Samples and methods

2.1. Digital samples

For both LBM and DIA approach, a digitized data of a porous rock is essential. The digital image can be obtained using 3D imaging device such as X-Ray Micro-CT [2-5]. X-Ray Micro-CT scanner has the same principle as the medical CT scanner. However, in medical CT scanner, the detector and the X-Ray source rotates while in Micro-CT, the object rotates. A digital image can also be produced from computer modelling. The advantage of computer modelling, one can easily adjust the required parameters accordingly, and various models of porous media can be generated, ranging from a simple capillary cylindrical tube to a complex structured sedimentary rock model.

In this study we first generate two sets of models to compare the result obtained from DIA and LBM. The first set are three cubes with sidelength of 150 pixel which have a cylindrical shaped pipe as the pore space at the center of the cube. The diameters of the pipes are 100, 70, and 40 pixel (see figure 1).

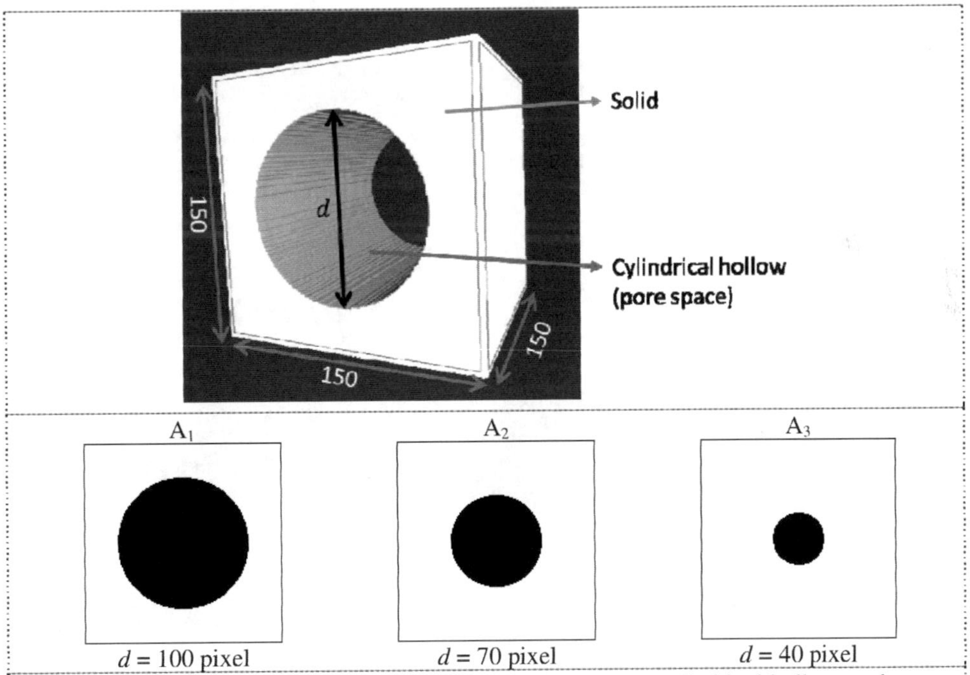

Figure 1. Top: the cube model of porous medium with simple cylindrical hollow as the pore space (white is the solid part, the gray colored area is the cylindrical pore wall). Bottom: 2D slice view of the sample with different diameter (white: solid, black: pore).

The second set are also three cubes with different sidelength which also have cylindrical shaped pipe as the pore space at the center of the cube. The sidelength of the cubes are varied as well as the diameter of the cylindrical pore as follows: 100:150, 90:60, and 60:40 (see figure 2). We also used the digital image of Fontainebleau sandstone to perform the permeability analysis. The digital sample of Fontainebleau sandstone has the size of 300×300×299 pixels (see figure 3), with spatial resolution of 7.5 μm/pixel. Fontainebleau sandstone is originated from France which has unique characteristics. It composed of 99% quartz, virtually devoid of clay, well recognized as a well-sorted, medium-grained, which sometimes denoted as a good example of clean sandstone. Fontainebleau sandstone is often used as a calibration experiments, usually in the calculation of porosity and permeability.

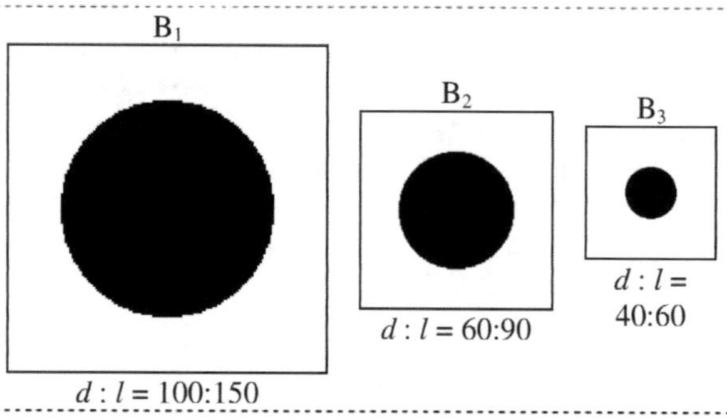

Figure 2. The second set of the model: three cubes which have the same sidelength:pore diameter of 2:3

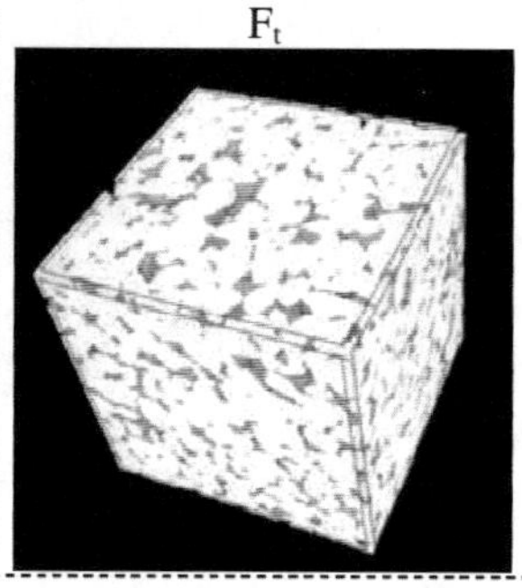

Figure 3. The third model – Fontainebleau Sandstone

2.2. Permeability
Permeability was first described quantitatively from an experiment carried out by Darcy in 1856 [6] which relates the pressure gradient of a fluid-saturated porous medium with the corresponding fluid flowrate as follows:

$$Q_x = -A \frac{\kappa}{\eta} \frac{\partial P}{\partial x}, \tag{1}$$

where Q_x is the volumetric fluid flowrate in x direction, κ is the permeability of the medium, η is the dynamic viscosity of the fluid, P is the fluid pressure, and A is the cross-sectional area perpendicular to the pressure gradient. LBM simulates Darcy's experiment by applying a pressure gradient to the inlet-outlet of a saturated porous medium. We use Palabos, a parallel Lattice Boltzmann solver (www.palabos.org) to perform the simulation [7]. Palabos utilized the D3Q19 scheme which describes motion in three dimensions and 19 associated velocity vectors. Permeability is then calculated using Darcy's law:

$$\kappa = -\frac{Q_x \eta}{A} \frac{1}{\partial P / \partial x}. \tag{2}$$

Volumetric fluid flowrate in a certain direction (in this case is x direction) can be expressed as the rate volume of fluid passing through the cross-sectional area A perpendicular to the flow direction as:

$$Q_x = \frac{V}{t} = \frac{Al}{t}, \tag{3}$$

where l is the length of the medium. Equation (2) can then be rewritten as:

$$\kappa = -\frac{Al}{t}\frac{\eta}{A}\frac{1}{\partial P/\partial x} = -\frac{l}{t}\eta\frac{1}{\partial P/\partial x}\ . \tag{4}$$

Equation (4) is practically modified to conform with Palabos as follows:

$$\kappa = \frac{\langle v\rangle \eta}{\partial P/\partial x}, \tag{5}$$

where $<v>$ denotes the average magnitude of the intrinsic velocity over the entire system volume.

Permeability has units of area (m^2 in SI units), but sometimes in terms of geophysical aspect, it is more common to use the Darcy where 1 Darcy = 0.987×10^{-12} m^2 or also millidarcy (mD) where 1 mD = 0.001 Darcy. There are some assumptions that are used in Darcy's law: only 1 phase of fluid exist in the pore, the fluid system is isothermal, the fluid is incompressible, the flow is laminar, the fluid is Newtonian, the fluid does not interact with the pore walls, and the fluid flow is steady state.

In this study, we also estimate the permeability using two empirical equations derived from the work of Kozeny-Carman. The equations contain general parameters which can be calculated from the digital image using image analysis. In this study we use CT Analyser software (Bruker-MicroCT) to calculate the general parameters such as specific surface area and porosity. The Kozeny-Carman equations are expressed as [8, 9]

$$\kappa = \frac{1}{2}\frac{\phi^3}{S^2\tau^2}, \tag{6}$$

$$\kappa = \frac{1}{8}\frac{\phi}{\tau^2}R^2, \tag{7}$$

where κ is permeability, ϕ is porosity, τ is tortuosity, S is specific surface area, and R is the average pore radius. Equation (6) and (7) are derived semi-analytically by comparing the flux in a pipe having circular cross-section with the Darcy's law [9]. We differentiate the symbol of permeability from each method as κ_{LBM} for permeability calculated using LBM, κ_{KC1} for permeability estimated using equation (6), and κ_{KC2} for permeability estimated using equation (7). Calculated permeability from equation (6) and (7) can also be expressed in lattice unit squared (lu^2) which can later be converted to m^2 if the spatial resolution of the digital data is available by multiplying the permeability in lu^2 with spatial resolution squared.

2.3. Error analysis

To evaluate the accuracy of the methods, we use analytical calculation of equation (6) and (7) and compare the result with the ones produced using DIA from the generated models. Analytical calculation can be easily done by using the predetermined parameters of the pore space i.e., the cylindrical pipe diameter and length, to which we can further analytically calculate porosity and specific surface area. The error, which is defined as the deviation of the result with regard to the analytical calculation is calculated using equation (8) for both κ_{KC1} and κ_{KC2}

$$\varepsilon = \frac{|\kappa_{KC,\mathrm{DIA}} - \kappa_{KC,\mathrm{ANA}}|}{\kappa_{KC,\mathrm{ANA}}}, \tag{8}$$

where $\kappa_{KC,\mathrm{DIA}}$ is permeability calculation using digital image analysis, $\kappa_{KC,\mathrm{ANA}}$ is permeability from analytical calculation for both equation (6) and (7) separately. Equation (8) is also used to calculate the error of the permeability calculated using LBM for dataset A and B by replacing $\kappa_{KC,\mathrm{DIA}}$ with κ_{LBM}. This definition of error is appropriate since $\kappa_{KC,\mathrm{ANA}}$ was actually derived semi-analytically for the cylindrical pipe model. As for the Fontainebleau sample, the errors are calculated by comparing to a known permeability of 1230 mD as calculated by Latief et.al. [10] as follows:

$$\varepsilon = \frac{|\kappa_{CALC} - \kappa_{REF}|}{\kappa_{REF}}, \tag{9}$$

where κ_{CALC} represents permeability obtained for the Fontainebleau sandstone by means of DIA using equation (6), equation (7), and from LBM (κ_{KC1}, κ_{KC2}, and κ_{LBM}).

3. Result and discussion

Table 1 shows the results from calculation on sample A_1, A_2, and A_3 and the corresponding calculated errors are listed in table 2. The permeability from analytical calculation ($\kappa_{KC,\mathrm{ANA}}$) using equation (6) and (7) yield the same results for all datasets A and datasets B since both equations are derived for the same geometrical sample.

Table 1. Result of permeability calculation on all the samples.

Methods		A_1 (lu^2)	A_2 (lu^2)	A_3 (lu^2)	B_1 (lu^2)	B_2 (lu^2)	B_3 (lu^2)	F_t (mD)
KC1	$\kappa_{KC1,\mathrm{ANA}}$	109.083	26.191	2.793	109.083	39.270	17.453	2318.470
	$\kappa_{KC1,\mathrm{DIA}}$	13.325	1.096	0.044	13.325	1.101	2.164	
KC2	$\kappa_{KC2,\mathrm{ANA}}$	109.083	26.191	2.793	109.083	39.270	17.453	4165.613
	$\kappa_{KC2,\mathrm{DIA}}$	109.257	26.239	2.824	109.340	34.192	17.761	
LBM	κ_{LBM}	108.683	26.071	2.791	108.683	38.984	17.446	1261.000

Table 2. First model error between LBM and DIA compared to Analytic

Methods	A_1 (%)	A_2 (%)	A_3 (%)	B_1 (%)	B_2 (%)	B_3 (%)	F_t (%)
KC1	87.78	95.82	98.42	87.78	92.12	87.60	88.49
KC2	0.16	0.18	1.11	0.24	12.94	1.76	238.67
LBM	0.37	0.46	0.07	0.37	0.73	0.04	2.52

The DIA approach for KC2 generally produce good results in contrary to the ones obtained using KC1 where the errors are ranged from 87% to 98%. The huge errors are caused by the characteristic of equation (6) which depends on three parameters compared to equation (7) which only depends on two parameters. The digital image analysis approach to estimate the parameters (ϕ, τ, S and R) might also cause the great difference of the calculated results. Porosity calculated from analytical and DIA generally do not differ much because the calculation using DIA uses simple summation. However, calculation of specific surface area and pore radius are more complex. Edge detection method for calculating the surface area produces overestimated value which later yield underestimated value of permeability if the image is pixelated. On the other hand, the estimated pore radius from DIA mostly do not differ much from the predefined analytical one.

From table 2 we can also observe that the LBM commonly produced consistent results which are similar to the ones calculated using analytical KC for datasets A and B. For the Fontainebleau (F_t) sample, the LBM also produced much better result compared to both KC equations. This simple analysis could give a straightforward conclusion, that LBM could provide us a more accurate result compared to the other two methods for both simple structure and complex structure. The drawback of implementing LBM to a complex structured porous media, is that the method could take hours to run, while using DIA, despite of its lack of accuracy, could visualize and characterize the sample in a matter of a few minutes. However, KC equations are derived semi-analytically which means the equations can still be modified accordingly depending on the applied medium structure. For example, a more generalized form of equation (6) can be used by replacing term ½ with a more general term $1/B$ where B is a geometric factor that accounts for the irregularities of pore shapes. In the case of sandstone, on can use value of B ranged from 2 – 5.

4. Conclusion and future works

From the performed analysis, we can draw several conclusions. The difference of the result from analytical calculation and estimation using DIA of the empirical equation (6) is larger than of the empirical equation (7). It is concluded that the equation (6) is not suitable for the model, however, equation (7) fits well with the generated models. Result from LBM and equation (7) are quite similar for the datasets A and B. As for the Fontainebleau sandstone, aside from LBM none of the KC equations are suitable for the sample due to the complexity of the pore structure of the porous medium. For the future works, a more reliable results from DIA could be provided by modifying the Kozeny-Carman equations with regard to the characteristic of the real sample.

5. Acknowledgment

This research was funded by the research program, “Riset Desentralisasi DIKTI-ITB 2015”.

References

[1] Filomena C M, Hornung J and Stollhofen H 2014 Assessing accuracy of gas-driven permeability measurements: a comparative study of diverse Hassler-cell and probe permeameter devices *Solid Earth* **5(1)** 1-11.

[2] Zacher G, M Halisch, P Westenberger, T Paul and T Tan 2014 Quantitative Investigation of Microstructures within Porous Rocks by Using Very High Resolution X-ray Micro-CT Imaging. *In EAGE/FESM Joint Regional Conference Petrophysics Meets Geoscience.*

[3] Atiko Hendriatna, Fourier Dzar Eljabbar Latief, Annisa Rachmi Trahwiwit Widijana, Fatkhan 2014 Digital Simulation to Predict Elastic Properties of Sandstone Samples Using Microtomographic Images, *Procs. PIT HAGI 2014*, 13 - 16 October 2014, Solo, Indonesia.

[4] Annisa Rachmi Trahwiwit Widijana, Fourier Dzar Eljabbar Latief, Almira Anissofia, Fatkhan, Handoyo 2014 Identification Cement by Using Digital Rock Imaging and Anaylsis Microscopic Image, *procs. PIT HAGI 2014*, 13 - 16 October 2014, Solo, Indonesia.

[5] Handoyo, Fatkhan, Fourier Dzar Eljabbar Latief 2014 Digital Rock Physics Application: Structure Parameters Characterization, Materials Identification, Fluid Modeling, and Elastic Properties Estimation of Saturated Sandstone, *procs. PIT HAGI 2014*, 13 - 16 October 2014, Solo, Indonesia.

[6] Darcy H 1856 *Les Fontaines Publiques de la Ville de Dijon,* Paris: Dalmont.

[7] Degruyter W, Burgisser A, Bachmann O, Malaspinas, Orestis 2010 Synchrotron X-ray Microtomography and Lattice Boltzmann Simulation of Gas Flow Through Volcanic Pumices. *Geosphere* **6** pp.470-471

[8] Carman PC 1961 *L'écoulement des gaz à travers les milieux*: University of France Press, Paris, France.

[9] Mavko G, Mukerji T, Dvorkin J, 2009 *The Rock Physics Handbook 2nd Edition.* Cambridge University Press, New York. United States of America.

[10] Latief FDE, Biswal B, Fauzi U, Hilfer R 2010 Continuum reconstruction of the pore scale microstructure for Fontainebleau sandstone, *Physica A*: Statistical Mechanics and its Applications **389 (8)** pp 1607-1618

Hydrostatic Simulation of Earth's Atmospheric Gas Using Multi-particle Collision Dynamics

Asis Pattisahusiwa[1], Acep Purqon[1], and Sparisoma Viridi[2]

[1] Physics of Earth and Complex Systems Research Division, Institut Teknologi Bandung, Indonesia.

[2] Nuclear physics and Biophysics Research Division, Department of Physics, Institut Teknologi Bandung, Indonesia.

E-mail: asis@pattisahusiwa.com

Abstract. Multi-particle collision dynamics (MPCD) is a mesoscopic simulation method to simulate fluid particle-like flows. MPCD has been widely used to simulate various problems in condensed matter. In this study, hydrostatic behavior of gas in the Earth's atmospheric layer is simulated by using MPCD method. The simulation is carried out by assuming the system under ideal state and is affected only by gravitational force. Gas particles are homogeneous and placed in 2D box. Interaction of the particles with the box is applied through implementation of boundary conditions (BC). Periodic BC is applied on the left and the right side, specular reflection on the top side, while bounce-back on the bottom side. Simulation program is executed in Arch Linux and running in notebook with processor Intel i5 @2700 MHz with 10 GB DDR3 RAM. The results show behaviors of the particles obey kinetic theory for ideal gas when gravitational acceleration value is proportional to the particle mass. Density distribution as a function of altitude also meets atmosphere's hydrostatic theory.

1. Introduction

Atmosphere is a layer that surrounds the earth and is mostly composed of molecular gases and plasma in upper atmosphere. Atmosphere is interesting to study since it has a lot of complex phenomena and still under active study. The complexity is mostly contributed by the interaction of fluids, such as the interaction between ocean and neutral gas, or between neutral gas and plasma. Hence, any computational fluid dynamics methods can be used to observe this complexity.

In the other hand, Multi-particle collision dynamics (MPCD) that was first proposed by Malevanets and Kapral in 1999 [1] offers a simple computational method to model fluids dynamics. In MPCD method, as well as in the other particle-based methods, the fluids are assumed as a collection of particle-like objects with zero volume. MPCD algorithm is similar to Direct Simulation Monte Carlo (DSMC) [2], the difference is only on the collision mechanism among the particles. The collision in DSMC method is represented by binary-collision between the pair of particles; while in MPCD, the collision is represented by stochastic rotation dynamics. MPCD with this collision mechanism is known as Stochastic Rotation Dynamics (SRD). MPCD with Anderson Thermostat (MPCD-AT) can be used as an alternative to SRD mechanism [3]. The method has been widely used to simulate various problems in the complex fluids. For example, simulation of solute-solvent dynamics [4, 5], claylike colloids [6], star polymers in solution [7], and flow-induced polymer translocation [8].

As far as we know, MPCD method has never been used in the atmospheric hydrodynamics or hydrostatics problems. So, the method will be tested for its ability when applied to atmospheric problems. For this purpose, MPCD will be applied to a simple problem, simulation of behavior of atmospheric neutral gas in hydrostatic equilibrium. The standard of hydrostatic atmospheric model used in this study will be described in the section 2, while details of the method and simulation mechanism will be explained in section 3.

2. Standard Atmospheric Model

Based on U.S. Standard Atmosphere 1976 [9, pp. 6–7], the atmosphere up to altitude 86 km is assumed to be dry, homogeneously mixed, and obeys ideal gas law. The effect of homogeneity leads to composition of the atmosphere up to this altitude have uniform molecular weight, $M_0 = 28.964\,43\,\mathrm{kg\,kmol^{-1}}$. The atmospheric composition is dominated by nitrogen gas N_2 with 78.084 %, oxygen gas (20.9476 %), and traces of Ar, CO_2, Ne, He, Kr, Xe, CH_4, and H_2.

Atmospheric hydrostatic state can be described by using ideal gas law. Supposed atmospheric temperature T, particle density ρ, and molecular weight M_0, the atmospheric pressure is given by

$$P = \frac{\rho R^* T}{M_0}. \tag{1}$$

Profile of particle density for any altitude can be deduced from equation (1) and is given by

$$\rho(z) = \rho_0 \exp\left(-\frac{(z - z_0)}{H}\right), \tag{2}$$

where z_0, z are initial and current altitudes, respectively. Scale height H is defined by

$$H = \frac{R^* T}{g},$$

where R^*, g are universal gas constant and gravitational acceleration respectively. If we assume temperature $T = T_0 = 288.15\,\mathrm{K}$, then the scale height is 8.4334 km.

3. Multi-particle Collision Dynamics

Simulation step in MPCD algorithm is divided into two steps, streaming and collision. In the streaming step, particle position is updated based on its velocity calculated in the collision step. The particle position for any time step is calculated by using

$$\vec{r}_i(t + \delta t) = \vec{r}_i(t) + \vec{v}_i(t)\delta t + \frac{1}{2}\vec{g}\delta t^2.$$

In this equation, i express the index of the particles, $\vec{r}(t + \delta t)$ is position at time $t + \delta t$, while $\vec{r}(t)$, $\vec{v}(t)$ are position and velocity at time t respectively.

In the collision step, particles are sorted into a collision cell with lateral size a. MPDC algorithm allows the particles to interact with each other in the same cell only. The interaction itself is represented by stochastic rotation matrix with random angle for every time step. Particle velocity after interaction is calculated by using

$$\vec{v}_i(t + \delta t) = \vec{u} + \mathbf{R}\delta\vec{v}, \qquad \delta\vec{v} = \vec{u} - \vec{v}_i(t),$$

where $\vec{v}$, $\vec{u}$, $\mathbf{R}$, $\delta\vec{v}$ are velocity after interaction, velocity of center of mass, rotation matrix, and relative velocity between particle and center of mass, respectively. The velocity of center of mass is expressed by

$$\vec{u} = \frac{1}{N_c}\sum_{i}^{N_c} \vec{v}_i(t),$$

where N_c is number of particle per cell. Two-dimensional rotation matrix is calculated by using

$$\mathbf{R} = \begin{bmatrix} \cos(n\theta) & -\sin(n\theta) \\ \sin(n\theta) & \cos(n\theta) \end{bmatrix},$$

where n is chosen between $+1$ and -1 randomly, while θ in this study is set to $90°$.

Interaction of the particles with the box is applied through the implementation of boundary conditions (BC). Periodic BC is applied on the left and the right of the box, while specular reflection and bounce-back is applied on the top and the bottom side respectively. Implementation of the last two BC is implemented by modifying tangential and normal components of particle velocity based on

$$\begin{aligned} \vec{v}_t &= (2\Gamma - 1)\,\vec{v}_t, \\ \vec{v}_n &= -\vec{v}_n, \\ \vec{v} &= \vec{v}_t + \vec{v}_n, \end{aligned}$$

where $\Gamma = 1$ for specular reflection and $\Gamma = 0$ for bounce-back.

In the equation (2), we can see that the density profile is dependent only on altitude. Hence, motion of the particles in two-dimensional coordinate is sufficient to achieve this profile. In this study, The simulation is performed by assuming that there are N particles with mass $m = 1$, position $\vec{r}$ and velocity $\vec{v}$. The particles are distributed randomly in two-dimensional box with size $10a \times 20a$, $a = 1$. The number of collision cells in the box is 200 and average number of particles per cell is fixed to 5. We set $R^*T = 1$, this is led to mean particle speed $v = 1.5958$, and speed of sound $c_s = 1.1832$.

Each particle is given a random initial velocity which follows normal distribution. The distribution is generated from Box-Müller transformation and expressed by [10]

$$\begin{aligned} y_1 &= \sqrt{-2\log(u_1)}\cos(2\pi u_2), \\ y_2 &= \sqrt{-2\log(u_1)}\sin(2\pi u_2), \end{aligned}$$

where u_1, u_2 are uniformly distributed random numbers, and y_1, y_2 are a pair of normal distributed random numbers. The pair of values are used interchangeably to produce the initial velocity for each particle through

$$\vec{v}_i = y\sigma + \mu; \qquad y = \begin{cases} y_1, & i \text{ even} \\ y_2, & i \text{ odd} \end{cases},$$

where σ and μ are standard deviation and mean of the distribution. In this study, $\mu = 0$, while σ is the thermal speed of particle.

Mean free path is set to $0.25a$ and corresponding to collision interval

$$\delta t = \lambda t_0; \qquad t_0 = a\sqrt{\frac{M_0}{R^*T}},$$

where t_0 is unit time. Mean free path is taken smaller than lateral size ($\lambda \ll a$) indicates that there are no molecular chaos and Galilean invariant [11]. This problem can be solved by using grid shifting method [11]. The method is implemented by multiplying the grid position before collision with any random number b and then moving back after collision. The random number is chosen in the range $+a/2$ and $-a/2$.

Since the system depends on value of gravitation acceleration, it will be varied in performing the simulation. The value is chosen among 1, 1.5, 2, 2.5, 3, and 3.5. The simulation itself is carried out until 5000 time steps. The density profile is generated from average density value from 2500 to 5000 time steps, while the particle speed distribution is taken from the last time step.

4. Results and Discussion

Simulation program is built by using Xojo (`http://www.xojo.com`) which based on BASIC language. The application is programmed and executed in Arch Linux with processor Intel i5 @2700 MHz and 10 GB DDR3 RAM. Figure 1a shows the average of simulation time for several total number of particles during both streaming and collision steps for each time step. Although, We can not compare this result with simulation time of others methods, MPCD is reliable as an alternative of lightweight method for fluid simulation. On the figure 1b, We can see that the algorithm satisfies energy conservation and therefore also fulfill linear momentum conservation.

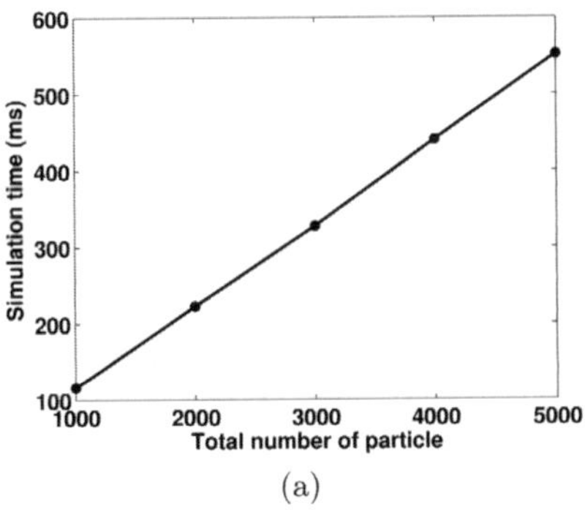

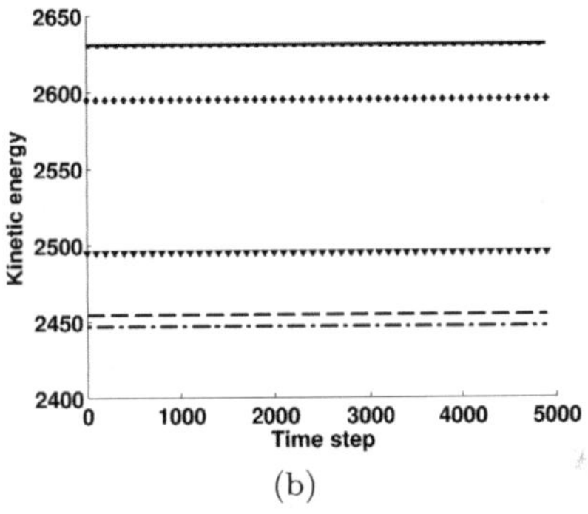

Figure 1: (a) Average simulation time for each total number of particles; (b) Profile of kinetic energy during simulation: $g = 1$ (dashed), $g = 1.5$ (solid), $g = 2$ (dotted), $g = 2.5$ (dash-dotted), $g = 3$ (▼), and $g = 3.5$ (♦).

Figure 2 shows that the system has achieved equilibrium when time step is close to 2000. This equilibrium is indicated by the trend of Mean Absolute Error (MAE) between density profile calculated from the model (2) and the simulation result. Figure 2 also shows that for small gravitational acceleration value ($g \leq 1.5$), fluctuations still occurred although the system has achieved equilibrium. Especially for $g = 1$ where the system is not stable enough although the trend is quite stable. In contrast to that, more stable fluctuations are shown by higher values of g.

According to Figure 3, MPCD algorithm is fairly accurate in modeling hydrostatic atmospheric system. The dash-dotted line on the figure is calculated from equation (2), while the other one is the simulation result. Hydrostatic density profile can be achieved when $g \geq 2$. However, when $g \geq 3.5$ it seems that the applied force is too big and attracts the majority of the particles to the bottom of the box, while for smaller g, particles are not really affected by applied force. These effects are also shown in the profile of particle speed distribution (Figure 4). Particle speeds do not obey Maxwell-Boltzmann distribution when gravitational acceleration is higher than 2.5.

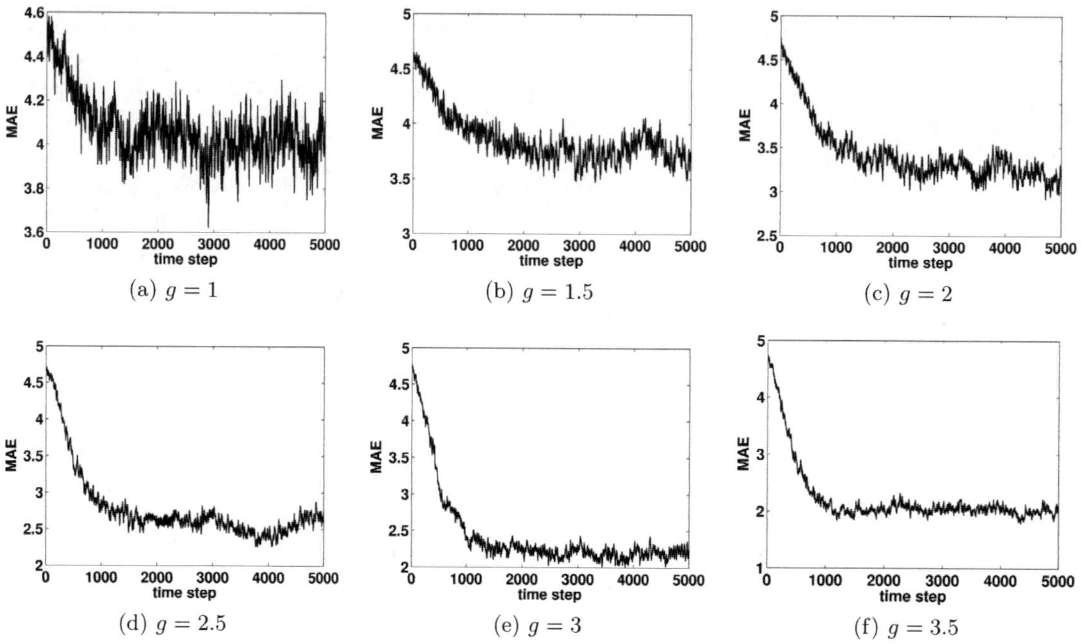

(a) $g = 1$ (b) $g = 1.5$ (c) $g = 2$

(d) $g = 2.5$ (e) $g = 3$ (f) $g = 3.5$

Figure 2: Mean Absolute Error (MAE).

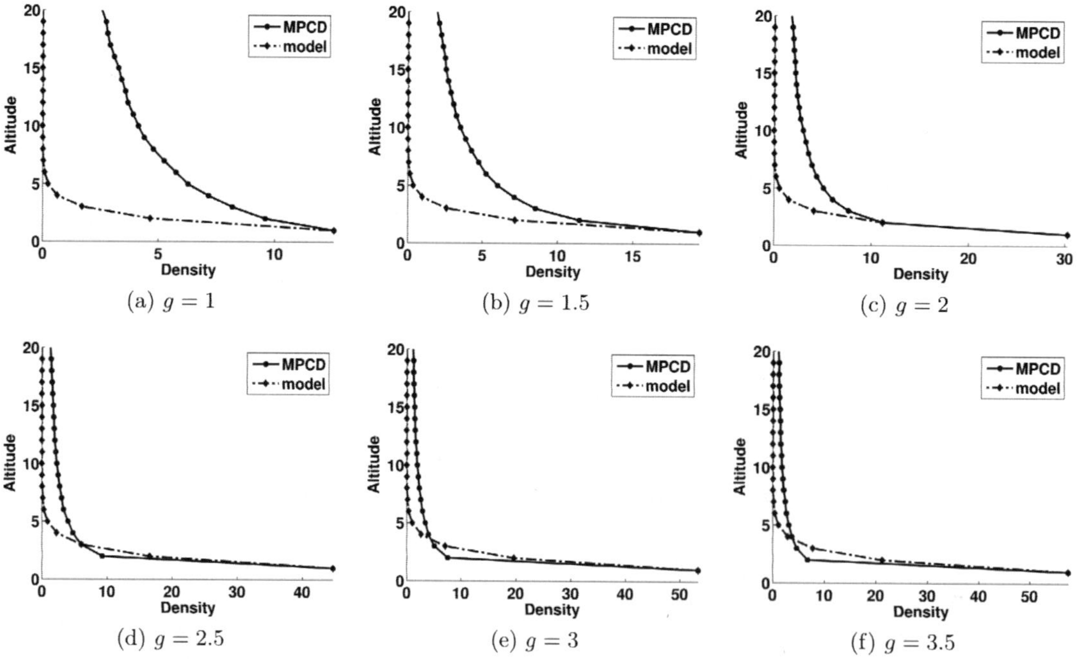

(a) $g = 1$ (b) $g = 1.5$ (c) $g = 2$

(d) $g = 2.5$ (e) $g = 3$ (f) $g = 3.5$

Figure 3: Average density profile as a function of altitude after equilibrium: Theoretical prediction (♦ dot-dashed line) and MPCD results (• solid line).

5. Conclusions

MPCD algorithm is quite fast and can be used as an alternative of lightweight (low performance computing) method for fluids simulation. Atmospheric hydrostatic equilibrium can be achieved

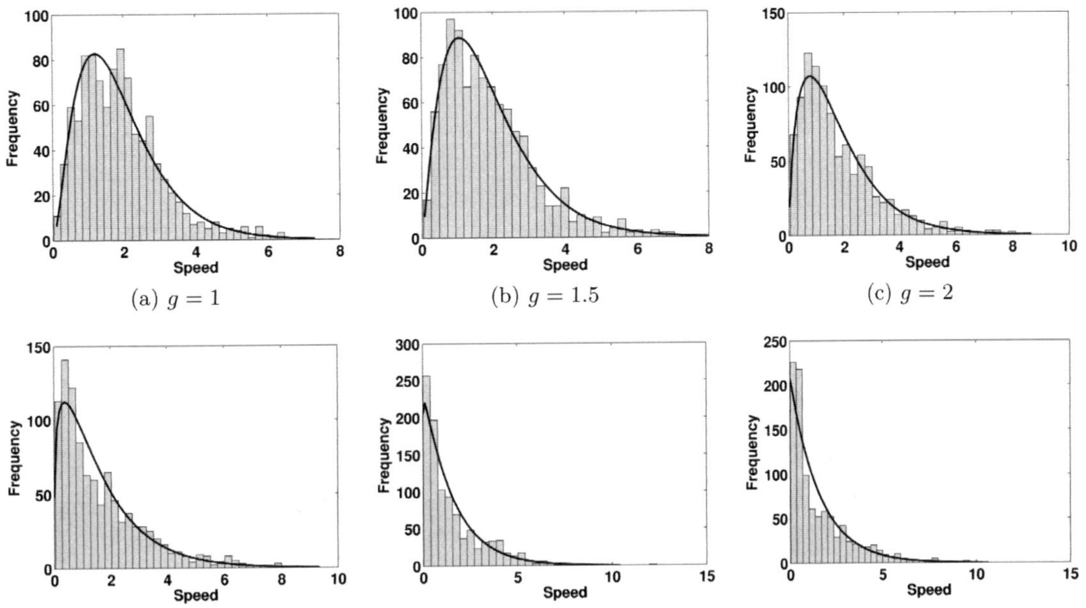

(a) $g = 1$ (b) $g = 1.5$ (c) $g = 2$

(d) $g = 2.5$ (e) $g = 3$ (f) $g = 3.5$

Figure 4: Particle speed distribution: MPCD result (bar) and fitted (solid line).

by using gravitational acceleration that is proportional to the particle mass. Based on the results, the recommended gravitational acceleration value is about $1.5 \leq g < 2.5$. MPCD method is capable to modeling atmospheric hydrostatic state and probably can be extended to hydrodynamic cases in atmospheric layer.

References

[1] Malevanets A and Kapral R 1999 *Journal of Chemical Physics* **110** 8605–8613

[2] Bird G A 1994 *Molecular gas dynamics and the direct simulation of gas flows* (Clarendon Press, New York)

[3] Gompper G, Ihle T, Kroll D M and Winkler R G 2009 Multi-particle collision dynamics: A particle-based mesoscale simulation approach to the hydrodynamics of complex fluids *Advanced Computer Simulation Approaches for Soft Matter Sciences III* (Springer Berlin Heidelberg) pp 1–87

[4] Malevanets A and Kapral R 2000 *The Journal of Chemical Physics* **112** 7260–7269

[5] Allahyarov E and Gompper G 2002 *Physical Review E* **66** 036702

[6] Hecht M, Harting J, Ihle T and Herrmann H J 2005 *Physical Review E* **72** 011408

[7] Singh S P, Huang C C, Westphal E, Gompper G and Winkler R G 2014 *The Journal of chemical physics* **141** 084901

[8] Nikoubashman A and Likos C N 2010 *The Journal of chemical physics* **133** 074901

[9] COESA 1976 *U. S. Standard Atmosphere, 1976* (U. S. Government Printing Office, Washington, D. C.)

[10] Lee D U, Villasenor J D, Luk W and Leong P H W 2006 *IEEE Transactions on Computers* **55** 659–671

[11] Ihle T and Kroll D M 2001 *Physical Review E* **63**

The two dimensional fold test in paleomagnetism using ipython notebook

Dedi Setiabudidaya[11] **and John D.A. Piper**[2]

[1] Department of Physics, Faculty of Mathematics and Natural Sciences, Sriwijaya University, Jln. Palembang-Prabumulih KM 32, Indonesia.
[2] Geomagnetism Laboratory, Department of Earth and Ocean Sciences, University of Liverpool, Liverpool L69 3BX, United Kingdom.

E-mail: dsetiabudidaya@mipa.unsri.ac.id

Abstract. One aspect of paleomagnetic analysis prone to controversy is the result of the fold test used to evaluate the age of a magnetisation component relative to the age of a structural event. Initially, the fold test was conducted by comparing the Fisherian precision parameter (k) to results from different limbs of a fold structure before and after tilt adjustment. To accommodate synfolding magnetisation, the tilt correction can be performed in stepwise fashion to both limbs simultaneously, here called one dimensional (1D) fold test. The two dimensional (2D) fold test described in this paper is carried out by applying stepwise tilt adjustment to each limb of the fold separately. The rationale for this is that tilts observed on contrasting limbs of deformed structure may not be synchronous or even belong to the same episode of deformation. A program for the procedure is presented here which generates two dimensional values of the k-parameter visually presented in contoured form. The use of ipython notebook enables this 2D fold test to be performed interactively and yield a more precise evaluation than the primitive 1D fold test.

1. Introduction

The paleomagnetic fold test can be a powerful means for establishing the age of a magnetisation in a rock relative to an episode of deformation. It relies on adjusting the paleomagnetic directions from a distributed set of locations for geologic tilt and resolving whether they converge or disperse. The test was first suggested by Graham [1] and an example showing that magnetisation can survive stresses produced during folding was illustrated in Keweenawan sediments by Du Bois [2]. The test was placed on a statistical basis by McElhinny [3]. The Fisherian precision parameter of the directions of magnetisation in situ (k_b) is compared with their precision following tilt adjustment (k_a) and the ratio k_a/k_b is compared with F-ratio tables with equal degrees of freedom ($2N$-1), where N is the number of directions of magnetisation being tested. McElhinny [3] tabulated 95% and 99% confidence limits from application of this fold test for a range of N values.

[1] To whom any correspondence should be addressed.

Although the simple test of McElhinny [3] is often still used for cursory analysis of magnetisations in deformed rocks it was shown by McFadden and Jones [4] that the test is invalid, most importantly because the population of magnetisations either before or after adjustment cannot be Fisherian. The authors conclude "The correct statistical procedure for applying the fold test is to determine whether the mean direction of a group of sites from one limb may be distinguished from the mean direction of a group of sites from another limb of the fold". They showed that the test of McElhinny is in general too stringent so a positive result will almost certainly reflect a magnetisation that has survived folding. Unfortunately the new test of McFadden and Jones required a complicated sampling scheme which included several sites with the same tilt corrections [5].

With advancing research in the paleomagnetism of deformed rocks, complex deformation settings were encountered in which the McFadden and Jones [4] analysis was found to be no longer appropriate. As a result for a while workers tended to revert to application of the McElhinny test on the basis that the results produced were conservative and therefore at least dependable. To improve this situation McFadden [6] developed a further test which now satisfactorily covers most geological situations and numerous examples of application and discussion of this now widely-applied test can be found elsewhere e.g. in [7, 8, 9].

The basic assumption of all fold tests is that the rocks being tested were folded synchronously so that each tilt correction must be corrected with the same percentage unfolding. This condition requires that the rocks are embraced by a single fold structure and it excludes the possibility that one limb may have been more deformed than the other. To address this issue the present study embraces the possibility that the rock was originally magnetized in any configuration. Instead of using a complex statistical analysis, we take a visual approach in handling the fold test by making use of k contoured graphs generating by the data after executing stepwise tilt adjustments to the paleomagnetic data from both limbs. However, for full analysis we require information on the expected inclination, in term of an inclination band which can often be supplied from another source such as the paleomagnetic inclination from undisturbed layers or an estimate from the regional apparent polar wander path for the study area. In our opinion, an accurate inclination frame becomes a necessary condition for performing a fold test especially in regions of complex deformation. For a positive 2D fold test the age of magnetisation predates the age of folding and this is indicated by the condition where the point of 100% unfolding becomes the center of the k contours and is covered by the expected inclination band. If the age of magnetisation postdates the age of folding, the graph will show this when the point of 0% unfolding (i.e. *in situ* data) becomes the center of the k contours and is covered by the expected inclination band. The other configurations of the graph otherwise indicate synfolding magnetisations.

To find the preferred mean directions, data falling within the expected inclination band and having k-parameters greater than 95% of k_{max}, here called the target directions, were analysed using a bootstrap technique by calling sub-program `pmag.di_boot` [10]. The target directions are then resampled up to 5000 and the mean direction sought. The obtained means do not however, differ very much from the arithmatic means of the target directions. So, in this 2D fold test we only take the arithmatic mean by averaging the mean directions fulfilling the imposed k-parameter and inclination criteria because the process is much simpler and faster by calling `object.mean()`.

This 2D fold test program makes use of python sub-programs (`pmag`) written by Prof. Tauxe of Scripps Institutes of Oceanography (SIO) [10]. Python is a free scripting modern computer language and has a simple and clean syntax. The 2D fold test program is carried out within the ipython notebook environment. The ipython notebook is also freely available and has cross platforms which can be run under Linux, Apple OS and Windows. It provides a modern computation environment where the user can work in a shell basis so corrections can be done interactively without compilation and can run processes as in traditional computing. By using this technology, researchers can look at data, test new ideas and evaluate their outcomes and of course much of their time will be invested in solving the problem rather than in writing the computer program. It is estimated that the ipython notebook has now been used by 500,000 – 1.5 million people [11,12].

2. Application Examples of the 2D Fold Test

2.1. An example from the Devonian ('Old Red Sandstone') sediments of South Wales, United Kingdom

The Llanstephan fold is located in Lower Devonian redbeds near the northern margin of the Late Palaeozoic Variscan fold belt in South Wales at 51.76 °N and 355.59 °E. This fold belt is useful for demonstrating the present method because the folds are typically asymmetric with northern limbs tending to be steeper than southern limbs. The Llanstephan fold structure is well exposed and ideally selected for performing a fold test with the aim of dating the magnetisation relative to regional deformation. Paleomagnetic directions from seven sites were taken from the southern limb and four sites from the northen limb and analysed using the 2D fold test. These paleomagnetic data were reported by Setiabudidaya [13] as part of a wider investigation of the magnetisation record in the Devonian "Old Red Sandstone" sediments at the northern margin of the Variscan Belt in Britain [14] and originally analysed using the FOLDCONT program written in fortran. The site mean data are listed in Table 1.

Table 1. Paleomagnetic directions and field orientations of (a) the north limb (b) and the south limb of the Llanstephan fold, South Wales [13]. 'Site' = site number, 'Dec' = declination and 'Inc' = inclination of the magnetisation, 'Azm'=azimuth and 'Dip'= tilt of formation. All figures are in degrees (except for site numbers).

(a)

Site	Dec	Inc	Azm	Dip
1	206.9	-13.5	330	54
2	209.1	-18.4	327	41
3	210.0	-12.8	324	32
4	197.7	-2.9	342	65
5	208.6	-3.4	349	45
6	203.1	-5.0	333	64
7	206.8	-12.8	340	40

(b)

Site	Dec	Inc	Azm	Dip
1	208.0	-6.3	166	25
2	209.4	3.6	168	48
3	208.8	1.2	190	10
4	210.4	-1.4	167	18

In the first shell (Table 2) modules `matplotlib.pyplot`, `pmag`, `numpy`, `pandas` and `itertools` are imported. The last line `%matplotlib inline` should be assigned for the resulting graph to be shown on the computer screen. Table 3 shows the module `pandas` reading the datafile in this example residing in folder `Data/SouthWales/Llanstephan` with the subfilenames `northlimb.txt` and `southlimb.txt` respectively. The last two lines extract the numeric values of the data. The next shell shown in Table 4 assigns the expected inclinations in terms of an inclination band with the center `i0`.

Table 2. Instructions in python for importing the modules used in the 2D fold test.

```
import matplotlib.pyplot as plt
import pmag
```

IOP Conf. Series: Earth and Environmental Science **29** (2016) 012007 doi:10.1088/1755-1315/29/1/012007

```
import numpy as np
import pandas as pd
import itertools
%matplotlib inline
```

Table 3. Instructions in python for reading the data.

```
data_n_limb = pd.read_csv('Data/SouthWales/Llanstephan/
              north_limb.txt',sep=r"\s+")
data_s_limb = pd.read_csv('Data/South Wales/Llanstephan/
              south_limb.txt',sep=r"\s+")
data_n_limb = data_n_limb[['Dec','Inc','Azm','Dip']].values
data_s_limb = data_s_limb[['Dec','Inc','Azm','Dip']].values
```

Table 4. Instructions in python for assigning the expected inclination (in degrees).

```
i0 = -7.5
imin = -10.0
imax = -5.0
```

The python codes listed in Table 5 perform the stepwise tilt correction. In this example we take an nstep value of 50 so that the program calculates the *k* parameter, mean declination and mean inclination for each 2% of step of untilting in addition to *in situ* (0% unfolding). Table 6 shows the python codes used to plot the contour diagram of the data from Table 5 together with relevant attributes.

Table 5. Codes in python used to generate 2D data for *k*-values, mean declinations and mean inclinations following stepwise tilt corrections.

```
nstep = 50
data_n_limb = data_n_limb.transpose()
data_s_limb = data_s_limb.transpose()
data_n_limb_temp = data_n_limb.copy()
data_s_limb_temp = data_s_limb.copy()
x = np.linspace(0,1,nstep+1)
k = np.zeros((nstep+1,nstep+1))
k1d=np.zeros(nstep+1)
inc = np.zeros((nstep+1, nstep+1))
dec = np.zeros((nstep+1, nstep+1))
for i in range(nstep+1):
    data_n_limb_temp[3]=data_n_limb[3]*x[i]
    for j in range(nstep+1):
        data_s_limb_temp[3]=data_s_limb[3]*x[j]
        data_n_limb_final=data_n_limb_temp.transpose()
        data_s_limb_final=data_s_limb_temp.transpose()
        data_final=np.array([item for item in
                   itertools.chain(data_n_limb_final,
                   data_s_limb_final)])
```

```
        data_untilt=pmag.dotilt_V(data_final)
        data_untilt=np.array(data_untilt)
        tilted_direction=data_untilt.transpose()
        tilted_mean=pmag.fisher_mean(tilted_direction)
        k[i,j]=tilted_mean.get('k')
        inc[i,j]=tilted_mean.get('inc')
        dec[i,j]=tilted_mean.get('dec')
        if i==j:
            k1d[i]=k[i,j]
```

Table 6. Codes in python used to plot contours of *k*, declination and inclination after executing instructions in the previous shell (Table 5).

```
fig = plt.figure(figsize=(8, 8), dpi=100)
X,Y= np.meshgrid(x,x)
CS1=plt.contour(X,Y,k,colors='g')
CS2=plt.contour(X,Y,inc, levels=[imin, i0,imax],
    colors='k')
CS3=plt.contour(X,Y,dec,colors='r')
for c in CS2.collections:
    c.set_dashes([(0, (5.0, 5.0))])
for c in CS3.collections:
    c.set_dashes([(0, (20.0, 10.0))])
plt.xlabel('Unfolding North Limb(x 100%)')
plt.ylabel('Unfolding South Limb(x 100%)')
plt.clabel(CS1, inline=1,fmt='%1.0f', fontsize=10)
plt.clabel(CS2, inline=1, fmt='%1.1f',fontsize=10)
plt.clabel(CS3, inline=1, fmt='%1.1f',fontsize=10)
plt.title('Contour of k- and Dec-values and Expected Mean
Inclination')
```

 IOP Publishing
doi:10.1088/1755-1315/29/1/012007

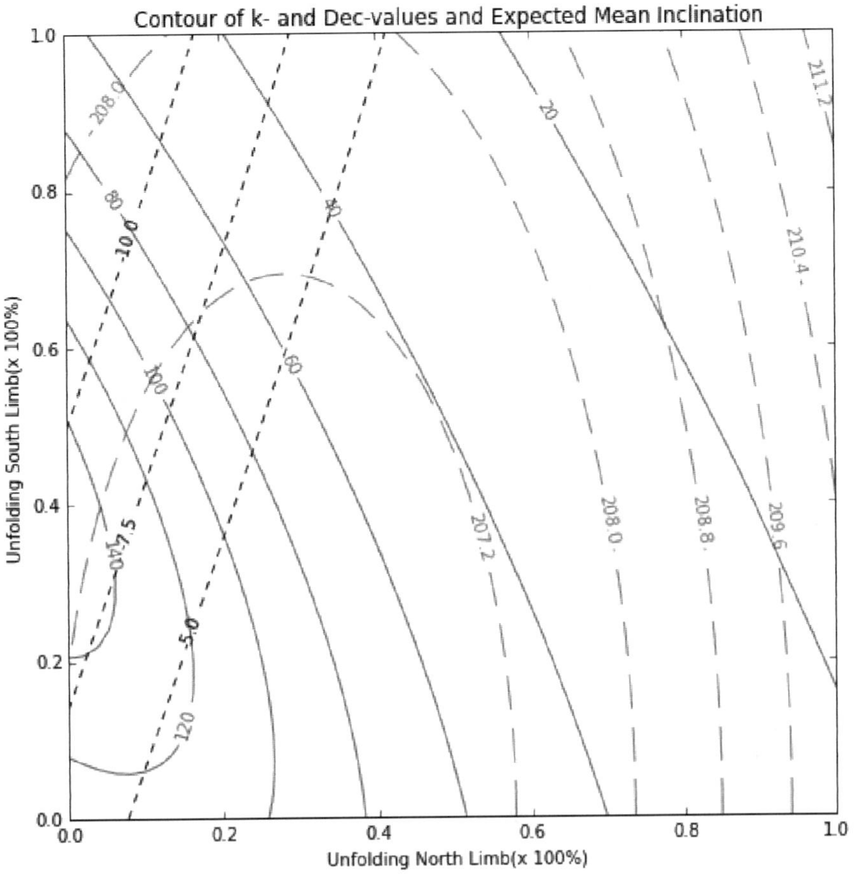

Figure 1. Contour diagram of *k*-parameters (solid green lines), three expected inclinations (short black dashed lines) and declinations (long red dashed lines) from the 2D fold test of the data shown in Table 1. Numbers shown are in degrees except *k* values.

The results illustrated in Figure 1 demonstrate that the age of magnetisation here is synfolding. The python codes used to find the mean paleomagnetic direction are listed in Table 7. Only data having *k*-values more than 95% of *k*-maximum and inclinations between -5.0° and -10° will be further analysed. In this example, there are 26 data conforming to these criteria. These 26 data have declinations ranging from 207.11° to 207.45°, inclinations ranging from -9.89° to -7.70° and *k*-values ranging from 143.39 to 150.77. These mean directions are obtained from configurations of 24% to 48% unfolding of the northern limb and 0% to 4% unfolding on the southern limb. For finding the mean direction a function mean is applied to each group of data according to the instructions listed in Table 8.

Table 7. Codes in python used to find the target directions with their corresponding k and unfolding configurations.

```
nstep=50
x = np.linspace(0,1,nstep+1)
k = np.zeros((nstep+1,nstep+1))
inc = np.zeros((nstep+1, nstep+1))
dec = np.zeros((nstep+1, nstep+1))
target_i = []
target_j =[]
target_dec = []
target_inc = []
target_k = []
for i in range(nstep+1):
    data_n_limb_temp[3]=data_n_limb[3]*x[i]
    for j in range(nstep+1):
        data_s_limb_temp[3]=data_s_limb[3]*x[j]
        data_n_limb_final=data_n_limb_temp.transpose()
        data_s_limb_final=data_s_limb_temp.transpose()
        data_final=np.array([item for item in itertools.
            Chain(data_n_limb_final, data_s_limb_final)])
        data_untilt=pmag.dotilt_V(data_final)
        data_untilt=np.array(data_untilt)
        tilted_direction=data_untilt.transpose()
        tilted_mean=pmag.fisher_mean(tilted_direction)
        k[i,j]=tilted_mean.get('k')
        inc[i,j]=tilted_mean.get('inc')
        dec[i,j]=tilted_mean.get('dec')
        if (k[i,j]>(0.95*kmax) and imax>inc[i,j]>imin):
        #we change the order of index i and j to fit the
        #graph ref. http://eli.thegreenplace.net/2014/
        #meshgrids-and-disambiguating-rows-and-columns-
        #from-cartesian-coordinates/
            target_i +=[(j*0.02)]
            target_j +=[(i*0.02)]
            target_dec +=[(dec[i,j])]
            target_inc +=[(inc[i,j])]
            target_k +=[(k[i,j])]
target_i=np.array(target_i);target_j=np.array(target_j);
target_dec=np.array(target_dec);
target_inc=np.array(target_inc);
target_k=np.array(target_k)
```

Table 8. Instructions in python for calculating the means.

```
print 'unfolding-x =',"%0.2f" %target_i.mean(),
      'unfolding-y =',"%0.2f" %target_j.mean()
print 'dec =',"%0.2f" %target_dec.mean(),'inc =',"%0.2f"
      %target_inc.mean(),'k =',"%0.2f" %target_k.mean()
print 'paleolatitude_2D =', "%0.2f",
      %pmag.plattarget_inc.mean())
```

Table 9. Output display of the result from applying the instructions of Table 8. The value of unfolding should be multiplied by 100%. The values of 'dec', 'inc' and 'paleolatitude_2D' are in degrees.

```
unfolding-x = 0.01 unfolding-y = 0.35
dec = 207.27 inc = -8.69 k = 146.76
paleolatitude_2D = -4.37
```

For a comparison, the mean direction from the 1D fold test is sought in terms of the maximum (1D) *k* value. The command and the printed output instructions are listed in Tables 10 and 11.

Table 10. Instructions in python for printing the results of the 2D fold test.

```
print 'unfolding =', "%0.2f" %(float(k1d.argmax())/nstep)
print 'dec =', "%0.2f" %dec[k1d.argmax(), k1d.argmax()],
      'inc =', "%0.2f" %inc[k1d.argmax(), k1d.argmax()],
      'k =', "%0.2f" %k1dmax
print 'paleolatitude_1D =', "%0.2f",
      %(pmag.plat(inc[k1d.argmax(),k1d.argmax()]))
```

Table 11. Output display of the 1D fold test. The value of unfolding should be multiplied by 100%. The values of 'dec', 'inc' and 'paleolatitude_1D' are in degrees.

```
unfolding = 0.10
dec = 206.84 inc = -5.24  k = 124.41
paleolatitude_1D = -2.62
```

The mean inclination derived from the 2D fold test is seen to be steeper than the mean inclination from the 1D fold test and places South Wales at a slightly more southerly paleolatitude during Permo-Carboniferous times than indicated by the result of the 1D fold test. The 2D fold test also predicts that the southern limb of the fold has experienced more stress than the northern limb since the formation was magnetized which is consistent with plate convergence from the south during the Variscan Orogeny and the frequent structural asymmetry observed in this region.

2.2. An example from Western Sumba, Indonesia

The Lasipu formation in Western Sumba is located at 9.7 °N and 119.6 °E. The formation is mildly deformed by open folding [15]. The azimuth information listed by the author assumed that the samples were taken from the same limb of a fold. Samples from site SL.D and SL.E revealed positive inclinations (normal) and samples from SL.F and SL.G yielded negative inclinations (reversed). Table 12 lists the paleomagnetic results where the reversed directions have been converted to equivalent normal polarity by adding 180° to the declinations and changing signs of the inclination. The expected inclination values for input are listed in Table 13.

Table 12. *In situ* paleomagnetic directions and orientation of (a) the normal limb and (b) the reversed limb of a fold structure at Lasipu, Western Sumba, studied by

Wensink [15]. The *in situ* direction was obtained by this author following untilting of the paleomagnetic directions by a minus dip in the azimuth direction. See also the caption of Table 1.

(a)

Site	Dec	Inc	Azm	Dip
SL.D	239.4	21.2	106	12.5
SL.E	236.1	10.0	112	40.0

(b)

Site	Dec	Inc	Azm	Dip
SL.F	245.0	40.3	149	26
SL.G	235.2	29.8	113	16

Table 13. Instructions in python for assigning the expected inclination (in degrees).

```
i0=32.5
imin=30.0
imax=35.0
```

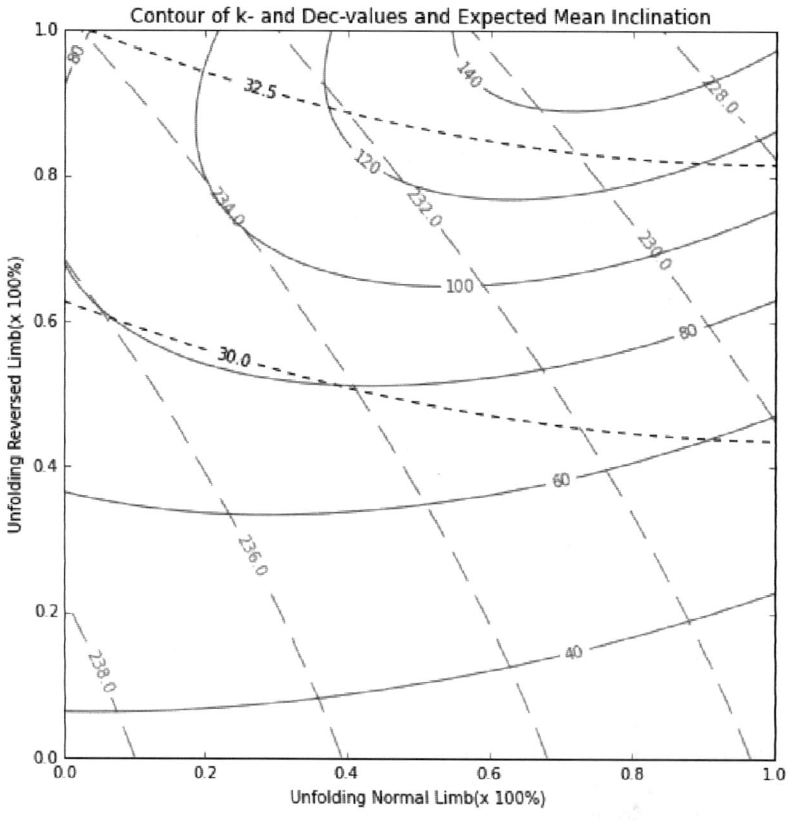

Figure 2. Contour figure of *k*-parameters (solid green lines), two expected inclinations (short black dashed lines (the third lies outside the range)) and declinations (long red dashed lines) from the 2D fold test applied to the data as listed in Table 2. Numbers shown are in degrees except *k* values.

The results of this test are shown in Figure 2. Using python code as in Table 7, we find that there are 51 data with $k > 95\%$ from k_{max} ranging from 146.69 to 154.19, the declination ranging from 227.13° to 229.65° and the inclination ranging from 33.14° to 33.58°. These mean directions are obtained from configurations of 94% up to 100% unfolding the normal limb and 62% up to 96% unfolding of the reversed limb. To find the mean, a function mean is applied to each group of data (cf. Table 9) and the results are listed in Table 14. For comparison the output from the 1D fold test is listed in Table 15. This 2D fold test shows that the predicted tectonic rotation of the study area is just 2° less than that predicted by the 1D fold test.

Table 14. Output from the 2D fold test: the value of unfolding should be multiplied by 100%. The values of 'dec', 'inc' and 'paleolatitude_2D' are in degrees.

```
unfolding-x = 0.78 unfolding-y = 0.98
dec = 228.63 inc = 33.40 k = 149.97
paleolatitude_2D = 18.24
```

Table 16. Output of the 1D fold test. The value of unfolding should be multiplied by 100%. The values of 'dec', 'inc' and 'paleolatitude_1D' are in degrees.

```
unfolding = 1.00
dec = 226.83 inc = 33.58  k = 143.98
paleolatitude_1D = 18.37
```

3. Conclusion

The 2D fold test, based on this pictorial approach shows more possibilities for adjusted mean paleomagnetic directions after executing stepwise tilt corrections than the 1D fold test. The expected inclination constraint is added together with a search for the highest k values. The mean paleomagnetic directions derived from the 2D fold test are more accurate and can therefore be used to revise paleomagnetic directions arising from the 1D fold tests. Hence, the paleolatitude and the amount of block rotation of the study area and the configuration of the limbs of a fold can collectively provide additional information to workers investigating the kinematics of complex deformed areas.

Acknowledgments

We thank Professor N.L. Swanson-Hysell of the University of California, Berkeley for assistance with the application of his program IPmag.py (http://nbviewer.ipython.org/github/ Swanson-Hysell/PmagPy_IPython/blob/master/IPmag_examples.ipynb). We also thank the PEDISGI 2015 organizer of Padjadjaran University, Indonesia for making this paper publicly available.

References

[1] Graham J W 1949 The stability and significance of magnetism in sedimentary rocks *J. Geophys. Res.* **54** 131-167
[2] DuBois P 1955 Palaeomagnetic measurements of the Keweenawan *Nature* **176** 500-507
[3] McElhinny M W 1964 Statistical significance of the fold test in palaeomagnetism *Geophys. J. R. Astr. Soc.* **8** 338 – 340
[4] McFadden P L and Jones D L 1981 The fold test in paleomagnetism *Geophys. J.R. astr. Soc.* **67** 53-58
[5] McFadden P L 1998 The fold test as an analytical tool *Geophys. J. Int.* **135** 329-338
[6] McFadden P L 1990 A new fold test for palaeomagentic studies *Geophys. J. Int.* **103**

163-169
[7] Watson G S and Enkin R J 1993 The fold test in paleomagnetism as a parameter estimation problem *Geophys. Res. Lett.* **20** 2135-2137
[8] Tauxe L and Watson G S 1994 The fold test: an eigen analysis *Earth planet. Sci. Lett.* **122** 331-341
[9] Weil A B and Van der Voo R 2002 The evolution of the paleomagnetic fold test as applied to complex geologic situations, illustrated by a case study from northern Spain *Physics and Chemistry of the Earth* **27** 1223-1235
[10] Tauxe L 2015 *Essential of Paleomagnetism* 3rd Web Edition http://earth.org/MAGIC /books/Tauxe/Essentials/
[11] Fernando P and Granger B E 2007 IPython: A System for Interactive Scientific Computing *Comput. Sci. Eng.* **9** 21-29
[12] Shen H 2014 Interactive Notebooks: Sharing the Code *Nature* **515** 151-152
[13] Setiabudidaya D 1991 *Magnetostratigraphy and Tectonic Rotation in South Wales* Unpublished PhD Thesis University of Liverpool (UK)
[14] Setiabudidaya D, Piper J D A and Shaw J 1994 Palaeomagnetism of the (early Devonian) Lower Old Red Sandstones of South Wales: implication to Variscan overprinting and differential regional rotations *Tectonophysics* **231** 257-280
[15] Wensink H 1994 Paleomagnetism of rocks from Sumba: tectonic implications since the late Cretaceous *Journal of Southeast Asian Earth Sciences* **9** 51-65

Geothermal well behaviour prediction after air compress stimulation using one-dimensional transient numerical modelling

W Yusman[1,4], **S Viridi**[2,4] and **S Rachmat**[3,4]
[1]Geothermal Engineering, Faculty of Mining and Petroleum Engineering
[2]Nuclear Physics and Biophysics, Faculty of Mathematics and Natural Science
[3]Petroleum Engineering, Faculty of Mining and Petroleum Engineering
[4]Institute Teknologi Bandung, Ganesha 10 st. Bandung 40132, Indonesia

E-mail: windrayusman@gmail.com

Abstract. The non-discharges geothermal wells have been a main problem in geothermal development stages and well discharge stimulation is required to initiate a flow. Air compress stimulation is one of the methods to trigger a fluid flow from the geothermal reservoir. The result of this process can be predicted by using by the Af / Ac method, but sometimes this method shows uncertainty result in several geothermal wells and also this prediction method does not take into account the flowing time of geothermal fluid to discharge after opening the well head. This paper presents a simulation of non-discharges well under air compress stimulation to predict well behavior and time process required. The component of this model consists of geothermal well data during heating-up process such as pressure, temperature and mass flow in the water column and main feed zone level. The one-dimensional transient numerical model is run based on the Single Fluid Volume Element (SFVE) method. According to the simulation result, the geothermal well behavior prediction after air compress stimulation will be valid under two specific circumstances, such as single phase fluid density between 1 – 28 kg/m^3 and above 28.5 kg/m^3. The first condition shows that successful well discharge and the last condition represent failed well discharge after air compress stimulation (only for two wells data). The comparison of ρ_f values between simulation and field observation shows the different result according to the success discharge well. Time required for flow to occur as observed in well head by using the SFVE method is different with the actual field condition. This model needs to improve by updating more geothermal well data and modified fluid phase condition inside the wellbore.

1. Introduction

Geothermal energy developments in Indonesia are constantly faced with several challenges such as exploration and engineering aspects. In engineering, the main goal of geothermal energy utilization is to have it well discharging but sometimes the geothermal wellbore do not show any fluid discharges. This problem is caused by the presence of water column in the wellbore which is standing above the major permeable zone [1] and to initiate flow, a well discharge stimulation is required. One of the effective methods to stimulate non-flowing geothermal well is an air compression method. Basically, this method boosts the energy in the water column by conductive heating at the certain depth. The liquid is depressed and it will increase the potential energy in the compressed gas [2]. Especially in air compression method, the prediction of "well behaviour" after stimulation process can be described by using the Af/Ac method [3]. This method calculates the ratio between two areas in static pressure and temperature profile (defined as Af and Ac) in order to find an association between successful and failed attempts of the well discharge capability after stimulation trial. Sometimes the result of air compress stimulation prediction using the Af / Ac method represents a different outcome than well discharges test trial [1]. The uncertainty result by using previous prediction might be minimized with numerical modelling which describes the geothermal well condition during stimulation process. The one-dimensional numerical modelling method which is used in this study is modified by using Single Fluid Volume Element method (SFVE) in siphon system [4]. The modified model will be adapted to geothermal well condition during air compress stimulation process. The simulation result of this condition model will be expected to predict geothermal well behaviour after air compress stimulation (based on static pressure and temperature profile data).

2. Basic Theory

In this section all theory involved in the geothermal well stimulation is listed and briefly explained, including geothermal wells principle, wells problem and wells stimulation. In this part also is described the simulation using modified Single Fluid Volume Element method (SFVE).

2.1. Geothermal wells principle and problem

The geothermal wells have the capability to produce heat from sub-surface. The main process of geothermal wells system are heat and mass transfer of geothermal fluid from the reservoir to the earth's surface [5]. Especially in liquid-dominated geothermal system, the produced fluid can flow under the following condition.

$$P_{\text{reservoir}} > P_{\text{hydrostatic}} + P_{\text{air inside the well}} \quad (1)$$

According to the Equation (1), in order to discharge, reservoir pressure must be greater than the sum of hydrostatic pressure of water column and the air pressure inside the well. Geothermal wells are divided into two types based on discharge capability such as artesian well and non-artesian well [1]. Artesian geothermal well refer to the natural geothermal fluid flow through well casing to the surface and fulfil Equation (1) condition after opening the wellhead valve. Conversely, the non-artesian well refer to the no discharge behaviour after opening the geothermal wellhead. It is caused by the shallow water column inside the wellbore. This type of wells located in the liquid dominated geothermal systems which have varying elevation and consequently the water column in each well show a different depth [1]. The presence of water column at the shallow level inside the geothermal well will increase the hydrostatic pressure.

2.2 Geothermal wells stimulation and previous prediction method

The problem of non-artesian geothermal wells can be solved by using several stimulation methods. The principle of those stimulation methods is to decrease hydrostatic pressure inside the wellbore by reducing the density or thickness of water column [1]. There are several methods of well stimulation for non-artesian type such as air (or gas) compression, gas lifting, and well-to-well two phase injection and all of these methods are usually used in liquid dominated geothermal systems. In particular, the air compression method boosts the energy in the water column by conductive heating at the certain depth to which the liquid is depressed [2]. The water column temperature is increased by this phenomena and it will change the fluid density of a water column (flashing). This condition will initiate a discharge mechanism [2]. All of the circumstances before air compress stimulation are shown in Figure 1 and after air pressure compress shown in Figure 2.

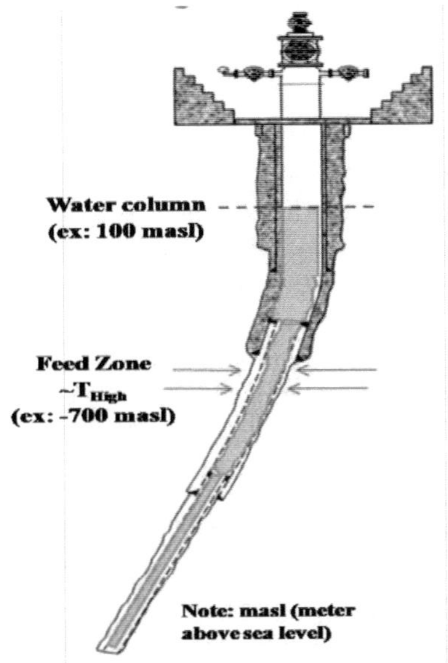

Figure 1. Initial condition of non-artesian geothermal wells.

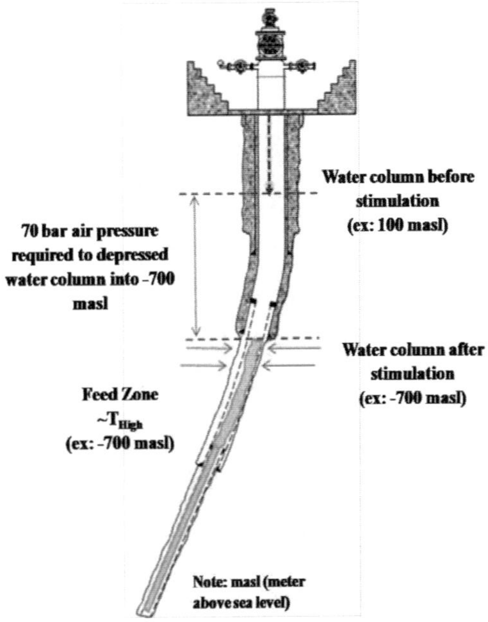

Figure 2. Final condition after air compress stimulation.

In air compression method, the successful prediction of the well discharge has been developed by a simple empirical formulation [3]. Basically, this formulation calculates the ratio between two areas such as Af and Ac. Af refers to the flashing area that is limited by the saturation temperature of the water column with wellbore temperature. Ac is the condensation areas including near the wellhead that is limited by the temperature curve in the well and the curve of 100 °C. The ratio between Af and Ac of some targeted wells were taken together with water column level before stimulation to find an association between successful and failed attempts, yet there are several limitations of this formulation. The first condition is an initial limiting case; the upper well temperature must be around 100 °C.

Table 1. The Af/Ac ratio probability range [3].

Af/Ac	Note
Af/Ac < 0.70	Low discharge probability
Af/Ac > 0.85	High discharge probability
Af/Ac = 0.70 – 0.85	Uncertainty range

Following to the Table 1, a ratio of Af and Ac less than 0.7 refer to the low discharge probability and the wells with Af/Ac above 0.85 define as high probability to discharge. A ratio between both values (0.75 to 0.85) is an uncertain result [3]. Sometimes, this well discharge prediction shows different result with a field data information on several geothermal wells.

2.3 Single Fluid Volume Element (SFVE)

The uncertain results by using the Af/Ac ratio can be minimized by developing a simple well model which represents the condition of non-artesian well before and after air compress stimulation. The wellbore model will be simulated based on well data, such as pressure, temperature and mass flow for each depth. One of the methods to represent the well stimulation process are one-dimensional transient flow simulation and this method related to physical processes in geothermal wellbore. This study is conducted by simulation of transient fluid flow in self-siphons using Single Fluid Volume Element Method (SFVE). Basically, this study can be modified to predict the flow occurrence in the geothermal systems.

SFVE method is a finite difference based method, which is used to solve equations of motion of a single fluid volume element. This element represents the transient motion of the fluid along its channel. Details of illustration and derivation of this method for vertical, horizontal, inclined, and semi-circular pipes can be found in [4], while only required equations are presented in this work. The element will have pressure force of fluid above and below it.

$$F_p = \rho g[(z + \Delta h) - z]A, \tag{2}$$

where A is well cross section, Δh thickness of the element, z position of the element, and ρ is the fluid density, which is in general

$$\rho(T,P) = \frac{\rho_0}{[1+\beta(T-T_0)][1-(p-p_0)/E]} \tag{3}$$

with ρ_0 = initial density (kg/m^3), E = Bulk modulus fluid elasticity (N/m^2) p_o = Initial pressure (N/m^2), β = Volumetric temperature expansion coefficient (m^3/m^3 °C), T_0 = Initial temperature (°C). In this study, β coefficient is assumed in the water condition (0.0002 m^3/m^3 °C at 20 °C) and the E coefficient is also assumed in the water condition (2.15×10^9 N/m^2) [8]. The initial density, pressure and temperature are taken in the specific depth. Temperature and pressure are also functions of depth z, which are provided by temperature- and pressure-depth profile. There is also force due pressure drop related to the position of the element relative to the initial position of shallow water column z_0.

$$F_d = -\rho g[(z - z_0)]A \tag{4}$$

Gravitation plays also a role of the element through

$$F_g = -\rho g A \Delta h \tag{5}$$

The last considered force is viscous drag between the element and its channel

$$F_v = -8\pi\eta\Delta h v \tag{6}$$

as in [4], which is modified from [6]. Parameter v is a velocity of fluid element and fluid viscosity η is also a function of temperature

$$\eta(T) = \eta_0 \times 10^{247.8/(T-140)} \tag{7}$$

with η_0= 2414 cP and T = fluid temperature (°K). Other forms [7] of Equation (7) can also be used. Then, using Newton second law of motion, acceleration of fluid element is

$$a = \frac{1}{\rho A \Delta h}\left(F_p + F_d + F_g + F_v\right) \tag{8}$$

Integration of Equation (7) will produce velocity, and integration will produce position. Numerical The method used is the simple Euler method.

$$v(t+\Delta t) = v(t) + a(t)\Delta t \tag{9}$$

and

$$z(t+\Delta t) = z(t) + v(t)\Delta t \tag{10}$$

Iteration though Equations (2) - (10) from an initial time to final time will give transient dynamics of a fluid element with thickness Δh. Based on simulation process, there are two condition to represent well behaviour in this study. The well behaviour is divided into two types such as discharge well and no discharge well after air compress stimulation. For simulation purpose, two termination conditions can be implemented, which $z > z_{\text{well head}}$ for the first type and $v < \varepsilon$ for the second type.

3. Geothermal wells properties

Data from two different wells are used in this work for testing the simulation, where "Well-A" represents unsuccessful well and "Wells-B" represents successful wells after air compress stimulation [1]. Figure 3 and Figure 4 show profiles of temperature-depth and pressure-depth of both wells.

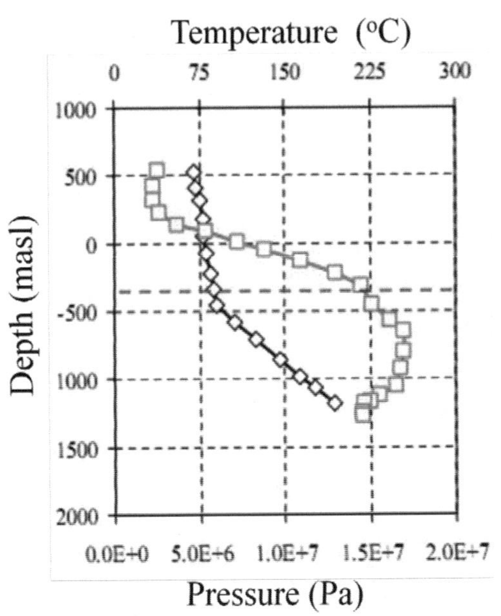

Figure 3. Pressure & Temperature static profile "Wells-A" before stimulation (failed attempt).

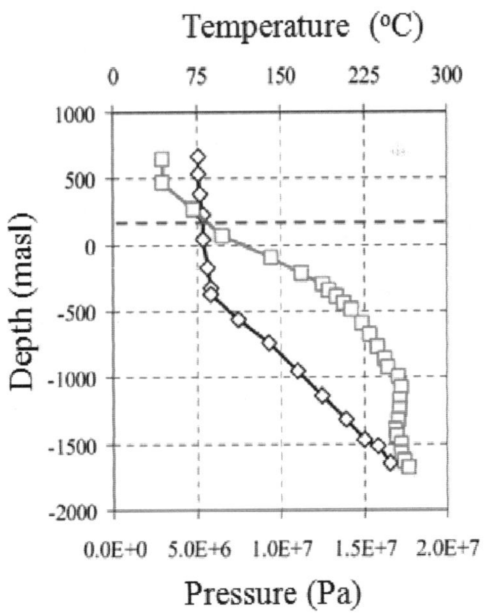

Figure 4. Pressure & Temperature static profile "Wells-B" after stimulation (successful attempt).

According to the Figure 3 and Figure 4, the pressure and temperature profile are represented by a blue and red line. The "Wells-A"has shallow water position at -350 m, temperature about 40 °C, pressure of 1 MPa, and it fails to discharge, where a "Wells-B" has shallow water position at about 200 m, temperature about 70 °C, pressure of 5.4 MPa, and it is successful to discharge.

4. Result & discussion

Simulation parameters in this study are Δh = 10 m and $\Delta t = 10^{-4}$ s. According to the simulation results which is given in Figure 5. It is found that the crossover between the flow and no flow states is located between values ρ_f = 28 kg/m^3 and ρ_f = 28.5 kg/m^3 for the well "Well-B", while for the "Well-A" all values of ρ_f give no flow. The ρ_f parameter refer to the final density of fluid element after simulation process ended. Following to the simulation result, the ρ_f value of single phase fluid also can be defined simply through

$$\rho_f = \alpha\rho_{water} + (1-\alpha)\rho_{vapor} \tag{11}$$

or

$$\alpha = \frac{\rho_f - \rho_{vapor}}{\rho_{water} - \rho_{vapor}} \tag{12}$$

In this study α is denoted as density fraction of water. The value of α is between 2×10^{-4} and 2.72×10^{-2} for the previous range of ρ_f simulation result. It is very vapour-rich mixture of water and vapour. For the success discharge well after air compress stimulation, the ρ_f can be measured in the geothermal well head. Based on "Well-B" observation data which is give a positive result after stimulation process, the ρ_f value is approximately 990 kg/m^3. It means that the simulation result using SFVE method (match observation only for values 1 kg/m^3 $<\rho_f <$ 28) is not approach the field data measurement. Time required for flow to occur as observed in well head is about 80 s simulation time, which is too short compared to real required time about more than 1000 second.

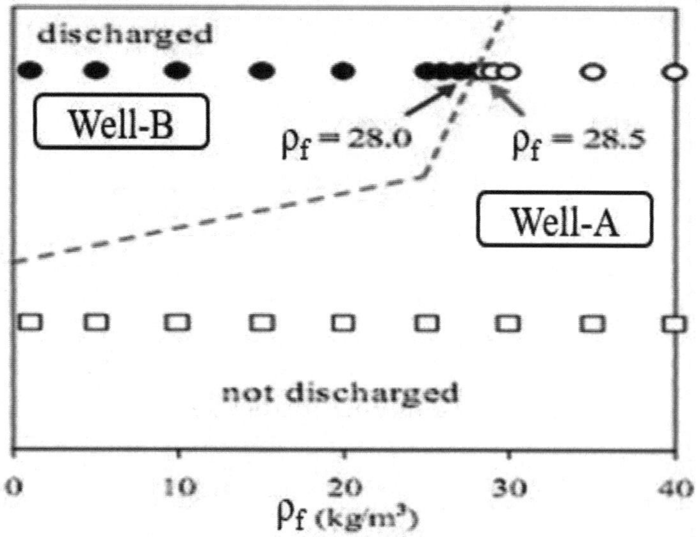

Figure 5. Flow occurrence as function of ρ_f for the two wells [1] using SFVE method.

This unrealistic result that the density of the discharged fluid is too light and discharge time is too short shows that implementation of SFVE, which considers a single-phase fluid, does not work well. The advancement of this method by considering two-phase flow. The unrealistic result can be reached due to the assumed static pressure-depth profile, which in reality changes during performing the SFVE.

5. Conclusion

The prediction of well behavior after air compress stimulation using 1-dimensional transient numerical modeling based on two geothermal wells will be valid with two specific circumstances. The first condition is single phase fluid density (ρ_f) between 1 – 28 kg/m^3 and the second condition is above 28.5 kg/m^3. Based on simulation result, the first condition shows successful well discharge and the second condition represents failed well discharge after air compress stimulation (only for two wells data). The comparison of ρ_f values between simulation and field observation shows the different result according to the success discharge well. Time required for flow to occur as observed in well head by using the SFVE method is different with the actual field condition. This model needs to be improves by modifying Single Fluid Volume Element (SFVE) method with two-phase flow assumption and collect more geothermal wells data with successful and failed discharge after stimulation.

Reference

[1] F. X. M. Sta. Ana 1985 *A Study on Stimulation by Air Compression on Some of the Philippine Geothermal Wells* Geothermal Institute. University of Auckland.
[2] Siega C, Saw V, AndrinoJr R and Canete G 2006 Well-toWell Two Phase Injection Using a 10in Diameter Line to Initiate Well Discharge in Mahagnadong Geothermal Field, Leyte, Philippines *Proc. of the 7th Asian Geothermal Symposium.*
[3] Stock 1983 *Condition Acquired for Successful Air Compression Stimulation* The Phillipines PNOC-EDC/KRTA Internal Report.
[4] Viridi S, Novitrian, Nurhayati, Latief F.D.E and Zen FP 2014 Development of Single Fluid Volume Element Method for Simulation of Transient Fluid Flow in Self-Siphons *AIP Conference Proceeding* **1615** 199-207.
[5] Grant M,A and Bixley P.F 2011 *Geothermal Reservoir Engineering* Academic Press Boston Elsevier Vol.2 pp. 269-283.
[6] Faber T E 1995 *Fluid Dynamic for Physicist* Cambridge: University Press pp 227-232
[7] Seeton C J 2006 *Viscosity –Temperature Correlation for Liquids* Tribology Letters Vol.1 pp 67-84.
[8] Density of Fluids – Changing Pressure and Temperature, http://www.engineeringtoolbox.com/fluid-density-temperature-pressured_309.html, Accessed: 2015 Mar 28th.

New Modified Band Limited Impedance (BLIMP) Inversion Method Using Envelope Attribute

Z L Maulana[1], O D Saputro, F D E Latief
Department of Physics, Faculty of Mathematics and Natural Sciences,
Institut Teknologi Bandung, Jalan Ganesa 10, Bandung 40132, Indonesia

E-mail: lazuardizulfikar@students.itb.ac.id.

Abstract. Earth attenuates high frequencies from seismic wavelet. Low frequency seismics cannot be obtained by low quality geophone. The low frequencies (0-10 Hz) that are not present in seismic data are important to obtain a good result in acoustic impedance (AI) inversion. AI is important to determine reservoir quality by converting AI to reservoir properties like porosity, permeability and water saturation. The low frequencies can be supplied from impedance log (AI logs), velocity analysis, and from the combination of both data. In this study, we propose that the low frequencies could be obtained from the envelope seismic attribute. This new proposed method is essentially a modified BLIMP (Band Limited Impedance) inversion method, in which the AI logs for BLIMP substituted with the envelope attribute. In low frequency domain (0-10 Hz), the envelope attribute produces high amplitude. This low frequency from the envelope attribute is utilized to replace low frequency from AI logs in BLIMP. Linear trend in this method is acquired from the AI logs. In this study, the method is applied on synthetic seismograms created from impedance log from well 'X'. The mean squared error from the modified BLIMP inversion is 2-4% for each trace (variation in error is caused by different normalization constant), lower than the conventional BLIMP inversion which produces error of 8%. The new method is also applied on Marmousi2 dataset and show promising result. The modified BLIMP inversion result from Marmousi2 by using one log AI is better than the one produced from the conventional method.

1. Introduction

Information regarding Acoustic Impedance (AI) is essential for hydrocarbon reservoir characterization. Hydrocarbon reservoir is source of oil and gas which preserved in porous rock. Many inversion methods have been developed in recent years to obtain AI from seismic trace. One of the simplest and widely used inversion methods is the recursive inversion. In recursive inversion, the derived equation from the wave reflection is modified and the method principally inverts acoustic impedance in terms of reflectivity coefficient [1]. Acoustic impedance pseudologs have been developed by Lavergne in 1975 [2]. In 1978, Waters developed inversion technique called SAIL (Seismic Approximate Impedance Log) [3], which derives impedance value from seismic data and low frequency information by stacking the velocities and well log. New impedance inversion scheme developed from SAIL called band limited impedance (BLIMP) was created by Ferguson and Margrave in 1996 [4], which used 2 Hz and 10 Hz (low frequency) data. BLIMP is a modification of SAIL, which also normalizes the

[1] To whom any correspondence should be addressed.

spectra of computed impedance log from seismic trace, thus having the same range of value as the real impedance log. In 2014, Wu [5] proposed a method that generated acoustic impedance inversion technique using envelope attribute called Seismic Envelope Inversion (EI) [5]. The method combines full waveform inversion (FWI) technique and the envelope attribute. Envelope attribute contains ultra-low frequency (ULF). FWI technique requires ULF to recover the long wavelength of the background structure while the envelope attribute is used as smoothing background or initial model for FWI. Wu applied the envelope attribute as ULF due to the highly-cost use of low frequency source. The result shows that ULF from envelope attribute is suitable to be applied for initial model in FWI and furthermore, it yields a good result. In this study, new inversion technique is proposed by combining BLIMP and envelope attribute. The idea is basically to obtain low frequency data from an alternate source such as Envelope Inversion (EI) does. Thus, we use envelope attribute as ULF source for BLIMP. In order to assess the performance of the method, we apply the method on a case from a single well. A numerical test using Marmousi model was also performed in order to verify the accuracy of the proposed inversion technique. Marmousi model is a subsurface model that generated by a consortium which based on a geological model in North Quenguela Through in the Quanza Basin in Angola [6, 7].

2. Theory

2.1. Band Limited Impedance (BLIMP)

BLIMP method starts by computing impedance which is estimated from seismic traces. BLIMP uses equation (1) to invert seismic trace into impedance. Equation (1) is similar to the recursive inversion but with some modifications and assumptions.

$$I_{j+1} = I_1 \exp(\gamma \sum_{k=1}^{j} S_k) . \tag{1}$$

Equation (1) integrates the seismic traces (S_k) and exponentiates the result to provide impedance traces. Variable γ is a scaling factor for the approximation in BLIMP. The equation is modified from Waters by including a scaling factor to estimate the impedance [3]. The SAIL method of Waters is summarized by Ferguson in the following procedure [4]:

i. Obtain an impedance log,
ii. Integrate then exponentiate each trace of seismic section,
iii. Compute Fourier spectra of (**i**) and (**ii**),
iv. Scale the spectra of impedance estimate from traces, so it macthes with the values in impedance log range,
v. High pass filter impedance log and add to the estimated impedance,
vi. Inverse Fourier transform of (**v**).

Ferguson and Margrave [4] modified the procedure by removing linear trend of impedance log before the Fourier transform, and adding linear trend of impedance log after inverse Fourier transform. The new modified technique is called BLIMP and is currently used widely in seismic inversion.

2.2. Envelope Attribute

Envelope attribute contains ultra-low frequency information of a seismic trace [5]. Seismic trace can be expressed as an analytic signal with real and imaginary component. In seismic acquisition, we only record the real part ($r(t)$) which is a combination of envelope amplitude ($A(t)$) and phase ($\theta(t)$). The real part ($r(t)$) and the imaginary part ($q(t)$) of seismic trace are defined as:

$$r(t) = A(t) \cos \theta(t) \tag{2}$$

$$q(t) = A(t)\sin\,\theta(t) \tag{3}$$

and the envelope attribute ($A(t)$) can be expressed as:

$$A(t) = \sqrt{r(t)^2 + q(t)^2}\;. \tag{4}$$

Imaginary part of seismic trace can be obtained from the real part using Hilbert transform. The Hilbert transform trace operator ($f(t)$) can be written in equation (5) and the imaginary part is expressed in equation (6).

$$f(t) = r(t) + iq(t) \tag{5}$$

$$q(t) = \mathrm{Im}\left[f(t)\right]. \tag{6}$$

3. Methodology

In this study, the inversion method is based on the idea of Envelope Inversion (EI) method. Low frequency signal from envelope attribute is able to recover the background structure. In this method, we propose to take account of low frequency from envelope attribute into BLIMP method. The following procedure describes the modified BLIMP method:

a. Compute the Envelope Attribute from each seismic trace,
b. Compute linear trend of (a) and subtract it,
c. Compute linear trend of AI log,
d. Compute the Fourier Spectra from (b),
e. Determine the scalar value between (d) and seismic spectra (normalized) and apply to (d),
f. Low pass filter (e) and add to seismic trace spectra,
g. Inverse Fourier Transform of (f),
h. Compute the Integrated Band Limited Impedance of seismic trace from equation (1), use I_1 only from one well and apply to all trace,
i. Add linear trend (c) to (h).

Note that the AI log is used to obtain the linear trend. However, we replaced the AI log in low frequency domain with envelope.

4. Result and Discussion

4.1. Analysis of Envelope Attribute

Envelope attribute has high amplitude in low frequency domain. A simple analysis of envelope attribute in synthetic trace is demonstrated as follows:
Ricker wavelet with dominant frequency of 15 Hz is used to produce a synthetic trace by convolving it with synthetic reflectivity coefficient. The produced synthetic seismic trace is shown in Figure 1. The trace has negative amplitude while the envelope attribute trace as shown in Figure 2 does not have negative amplitude. The frequency domain of the synthetic seismic trace is shown in Figure 3. We can observe in Figure 3 that frequency domain of the seismic trace shows low amplitude of the low frequency signals. However, the frequency domain of the envelope attribute (Figure 4) shows very high amplitude of the signal with frequency in the range of 0-10 Hz. We can observe that the trace does not contain low frequency, on the other hand, the frequency domain of the envelope attribute shows large quantity of low frequency signal in the range of 0-10 Hz.

A better comparison can be seen in Figure 5 where the frequency domains of the two traces are plotted in the same axis. The dominant frequency of the wavelet is increased to 50 Hz in Figure 6 and we can observe a significant difference in the low frequency area. The difference in 0-15 Hz frequency domain clearly shows that envelope attribute could produce a high amplitude in low frequency and thus considered very suitable to be applied in the modified impedance inversion method.

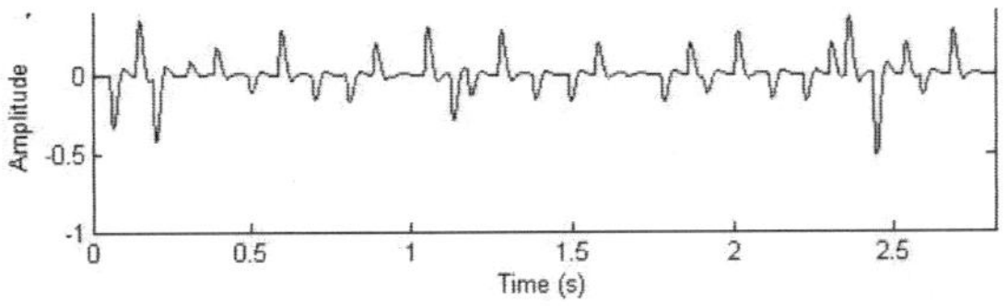

Figure 1. Synthetic trace with wavelet 15 Hz in time domain

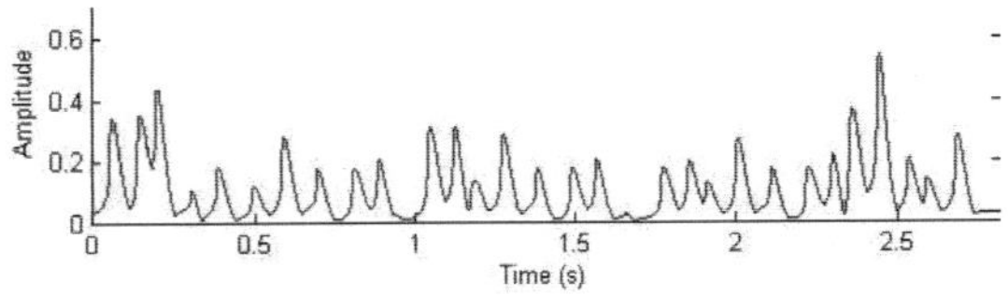

Figure 2. Envelope attribute from trace in time domain

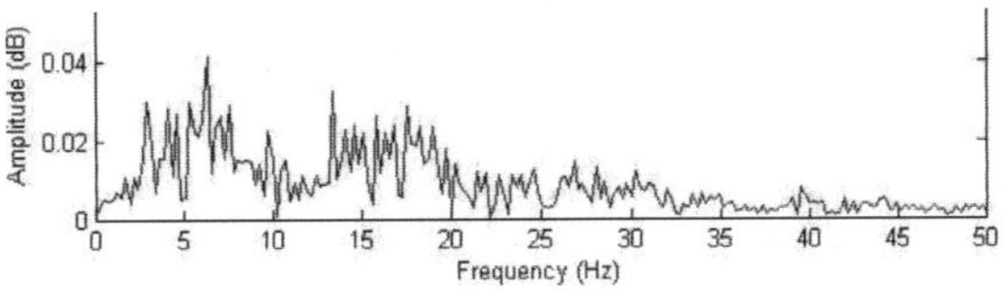

Figure 3. Synthetic trace with wavelet 15 Hz in frequency domain

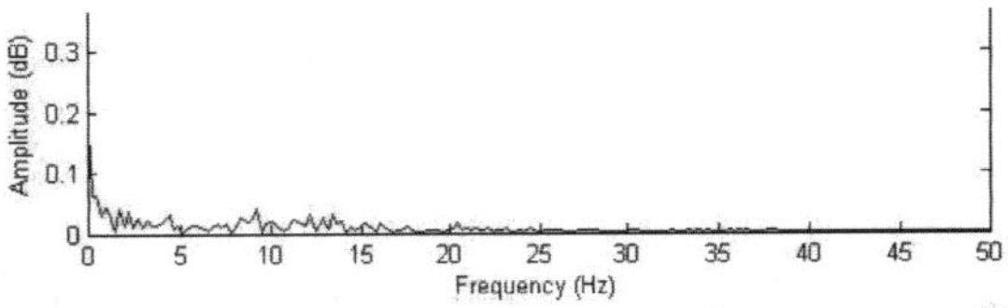

Figure 4. Envelope attribute from trace in frequency domain

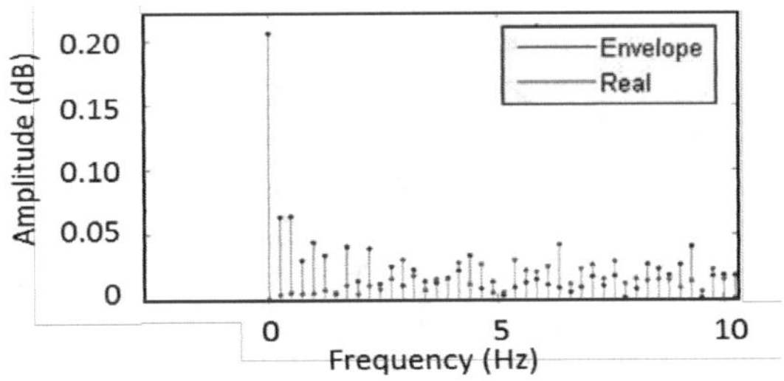

Figure 5. Envelope attribute (blue) show higher amplitude than seismic trace (green) with wavelet 15 Hz in low frequency domain.

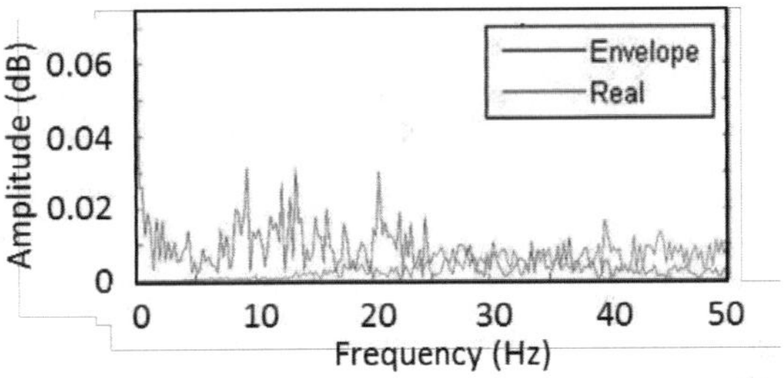

Figure 6. Envelope attribute (blue) show higher amplitude than seismic trace (green) with wavelet 50 Hz in low frequency domain.

4.2. Inversion test in field data

As an assessment of the proposed modified BLIMP method, we apply the procedure to a well log data from 'X' field. We compare the results from different AI inversion techniques, i.e., the recursive inversion, BLIMP, and the modified BLIMP. In Figure 7, the result from the three methods are compared. Recursive inversion, BLIMP and modified BLIMP has misfit value of 11.82 %, 8.1250% and 3.4% respectively. The proposed modified BLIMP has the smallest misfit value. We also found that the misfit percentage is good if the frequency cut off for envelope is in the range of 6-9 Hz as shown in Figure 8.

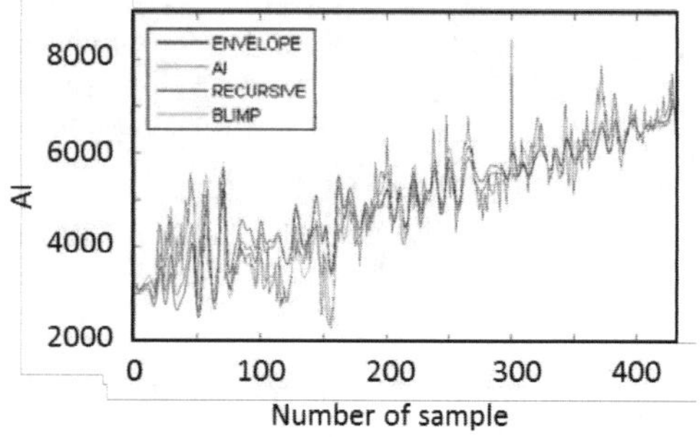

Figure 7. Inversion using modified BLIMP (blue), recursive plus AI log trend (red) and data log (green).

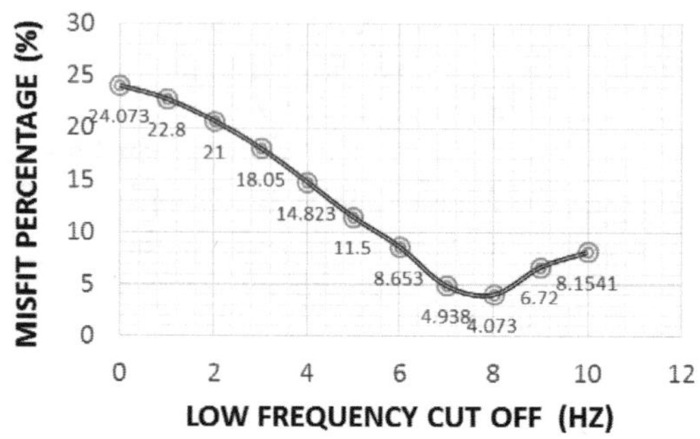

Figure 8. Misfit percentage in different low frequency cut off

4.3. Inversion test using Marmousi2

Marmousi model is a complex synthetic 2D acoustic data set which generated from the estimation based on a real situation. It contains many reflectors, steep dips and strong velocity gradients. Marmousi model was generated based on the North Quenguela Trough in Angola [6]. Original Marmousi model only provides the propagation of compressional waves. Marmousi2 model is a modified Marmousi model which supports not only compressional waves, but also shear waves, converted waves, and all kind of guided waves including those traveling along the sea floor [7]. The structure of Marmousi2 model is shown in Figure 9.

Inversion result of the Marmousi can be seen in Figure 9-12 where the color indicates the value of the relative acoustic impedance (AI), the horizontal axis is offset line and the vertical axis is depth. In Figure 9 we clearly see the detailed structure of Marmousi model from the acoustic impedance. In Figure 10, the recursive inversion expose structure contrasts such as fault and dip, however it cannot reveal any background structure color (which reflects the lithology). BLIMP method in Figure 11 shows background structure (colored) and thus we can determine the lithology. Although the result shows similar gradation of general structure, it fails to display the detailed structure. The Modified BLIMP inversion in Figure 12 produces adequate and even good detailed structure background compared to true Marmousi2 model in Figure 9. In Figure 12, the result from the modified BLIMP reveals structure contrast as well as the background structure (colored) that agrees satisfactory with the structure contrast. Compared to other inversion methods, modified BLIMP produces the best result.

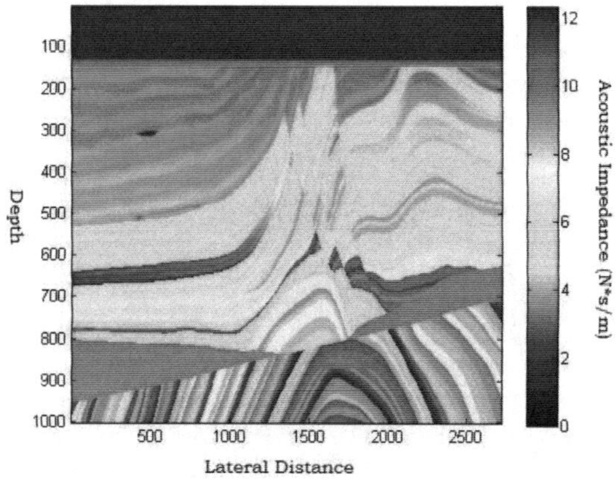

Figure 9. True Marmousi model (Color key in Kg*m/s^2)

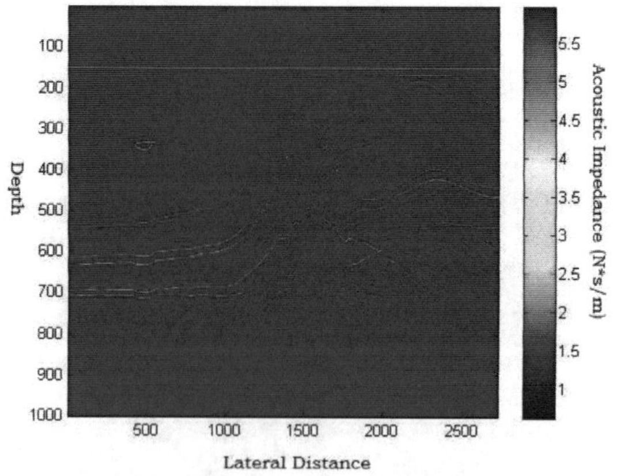

Figure 10. Inversion test using recursive inversion (Color key in Kg*m/s^2)

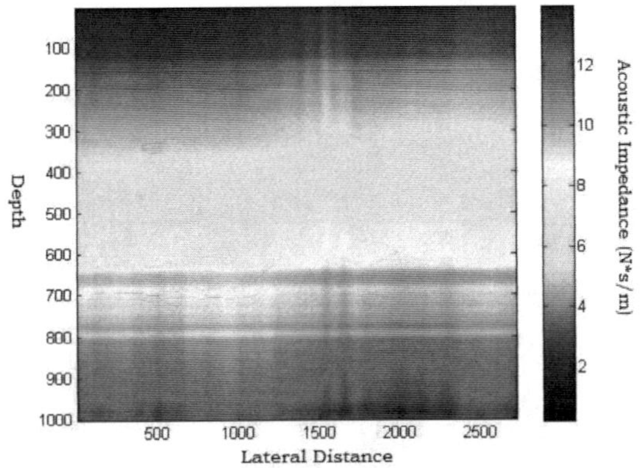

Figure 11. Inversion test using BLIMP (Color key in Kg*m/s^2)

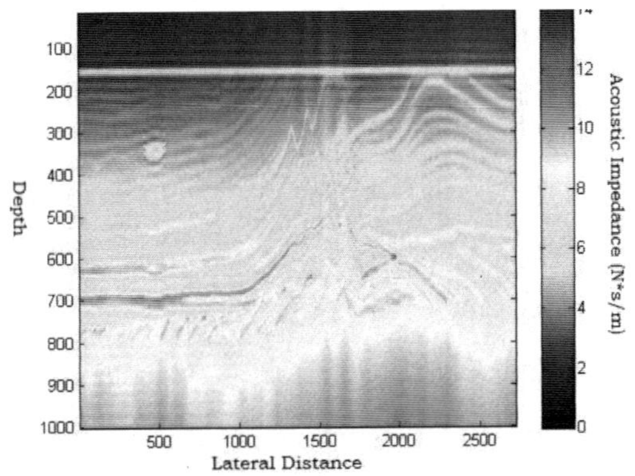

Figure 12. Inversion test using modified BLIMP (Color key in Kg*m/s^2)

Note that the comparison of true model and modified BLIMP inversion result (black circle) fails to identify the gas area. The detailed acoustic impedance in gas area is shown in Figure 13. The envelope attribute contributes to this error due to the nature of its equation (squared, thus impossible to reveal either negative impedance contrast, or positive impedance contrast).

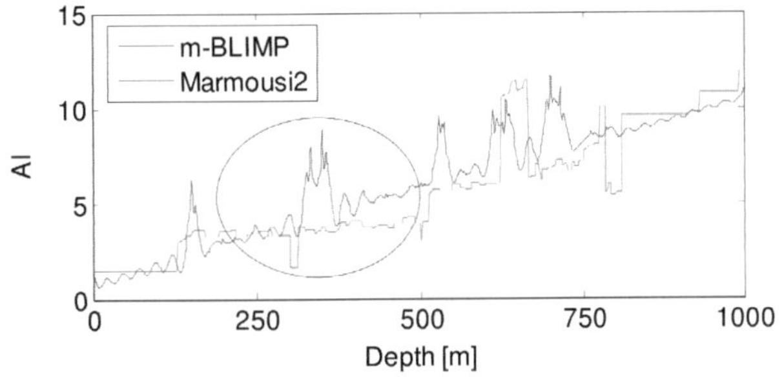

Figure 13. Acoustic impedance in gas saturated area for True Marmousi

5. Conclusion

The test from a well log data shows promising results which describe the capability of the proposed new modified BLIMP to produce more accurate AI inversion. The Modified BLIMP has better misfit than the recursive or the conventional BLIMP and has small misfit when applied to the 6-9 Hz cut off frequency. The modified BLIMP also yields the best result for Marmousi2 data, compared to recursive inversion and BLIMP. There is still a drawback of this method however, that the modified BLIMP cannot show the gas saturated area. One factor that distinguishes the modified BLIMP from other methods is that the modified BLIMP inversion can be carried out without having to build the velocity model or horizon picking, which is clearly very useful and potential in reducing cost of inversion.

Acknowledgement

We would like to thank Mr. Sonny Winardhie and Mr. Khairu Ummah for the valuable discussions and helps. The suggestions and comments of reviewers are appreciated. This research is partly supported by the research program "Desentralisasi DIKTI 2015".

Reference

[1] Brian Russel 1998, Introduction to Seismic Inversion Methods, *Society Exploration Geophysicists*, Tulsa
[2] M. Becquey, M. Lavergne, and C. Willm 1979 *J. Geophysics* **44** 1485-1501
[3] K. H. Waters 1978 Reflection seismology *John Wiley and Sons* New York
[4] Robert J. Ferguson and Gary F. Margrave 1996 A simple algorithm for band-limited impedance inversion *CREWES Research Report* 8 21
[5] Ru-Shan Wu, Jingrui Luo, and Bangyu Wu 2014 *J. Geophysics* **79** WA13-WA24
[6] Roelof Versteeg 1994 The Marmousi experience: Velocity model determination on a synthetic complex data set *The Leading Edge* **13(**9) 927-936
[7] Gary S.Martin, Robert Wiley, and Kurt J. Marfurt 2006 Marmousi2: An elastic upgrade for Marmousi *The Leading Edge* **25** 156-166

Body edge delineation in 2D DC resistivity imaging using differential method

Kusnahadi Susanto[1], Mohammad Fitrah Bahari[2]

[1]Geophysics Program Study, Universitas Padjadjaran, Indonesia
[2]Physics Program Study, Universitas Padjadjaran, Indonesia

k.susanto@geophys.unpad.ac.id

Abstract. DC resistivity is widely used to identify the kind of rock and the lithology contact. However, the image resulting from resistivity processing is shown in a contour image. There is be a problem to interpret where the edge of body location is. This study uses differential method to delineate the edge of body in DC resistivity contour. This method was applied to the boundary between gravel and underlying clay layer. The first and the second order differential method is applied to the delineation of lithology contact. The profiling curve has to be sliced and extracted from the resistivity contour before the differential method can be used. The spectral analysis shows the frequency and wavenumber of the profiling curve used to make gridding. The slicing process was conducted horizontally and vertically in order to get the mesh size which will be used in the differential method. The second order differential, the Laplace operator, is able to show the edge of body more clearly than the first order differential and shows the contact between gravel and clay.

1. Introduction

The geo-electrical 2D resistivity technique is widely used to investigate natural resources, archeology, geo-technique [1], hydrogeology and several environment problem [2,3]. The result of geo-electrical 2D resistivity imaging is usually shown as a contour map. It contains not only the information of lithology interpreted from resistivity value but also the information of the shape of the resistive body interpreted from shape of contour. Ambiguity often occurs when the contact of body lithology is to be determined. It is due to the degradation of color or limitations of contour interval on cross-section of 2D geo-electric. Therefore, the interval of contour should be set in order to reduce noise. Furthermore, we are able to focus on the investigated object. Many techniques have been developed to reduce the noise and to erase the outlier on the resistivity data [4,5]. Solving the ambiguity problem by numeric and simulation processing is very important and helpful to determine a resistivity model for the electrical imaging survey of the subsurface [6].

2. Method

The second order differential method or the Laplace operator method was used to analyze the contact of the body resistive on the result of 2D geo-electric imaging. This method uses a simple calculation to detect the body edge and is widely used in image processing for the edge detection technique. As we know, the data of 2D DC resistivity measurements has weakness in interpretation of the body edge. Figure 1 shows DC resistivity contour image in gravel-clay layering [4,7].

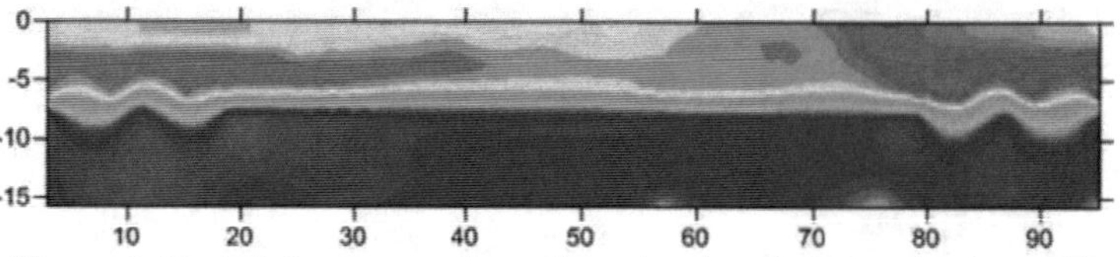

Figure 1. Resistivity contour from Gravel and underlying clay layer [8]

This technique combines differential method and the spectral analysis. The differential method was used to detect edge of body resistive while the spectral analysis was used to optimize computation process. Variable of Δx and Δy are used on Laplace operation to calculate the crossing zero, then the spectral analysis is able to give the optimal value based on the resistivity profile which is made to extract the wave number using Fourier transform and to calculate the mesh sizes Δx and Δy.

2.1 Spectral analysis
The calculation of spatial mesh is begun by slicing the contour lines of 2D DC resistivity data with x and y direction, where the profile attribute on vertical axis is resistivity (Ωm) and on horizontal axis is length or depth (m). The resistivity profile in spatial domain is processed by using Fast Fourier Transform (FFT) to separate the high and low frequency. The profile of high frequency requires a small grid size of Δx or Δy and the computational process will be costly while the profile of very low frequency creates large grid size and the resolution process become coarse. The cut off frequency is calculated to find the optimal point between high and low frequency, it is written as:

$$\rho'(f) = \int_{-\infty}^{\infty} \rho(x) e^{-j2\pi f x} dx \qquad (1)$$

where $\rho'(f)$is the resistivity frequency profile, $\rho(x)$ is the resistivity profile, f is frequency, x is the length or depth. In this way, the resistivity profile has to be made perpendicular to contour line. It will show slope on graphic and have multiple frequency on it. Zero frequency is caused by homogenous resistivity. The cut off frequency was extracted from resistivity frequency profile using FFT. The technique is the same as residual-regional separation on gravity method.

2.2 Differential Method
Pierre de Fermat has developed a differential algorithm to define the maximum, minimum and slope value of a special function. He used the first order differential as a technique to calculate the gradient of a curve. Figure 2 presents the gradient of a curve as the result of the first order differential. The direction of gradient line is written in the gradient value. In the process of the edge detection, the maximum and minimum values of the original function are not as important as the slope which presents the gradation contour that has the edge information. The first order differential generates the maximum and minimum values at the slope, then it results various values if the slope has multi-gradient so that it can cause confusing interpretation. The second order differential or the Laplace operator could be used to resolve that problem.

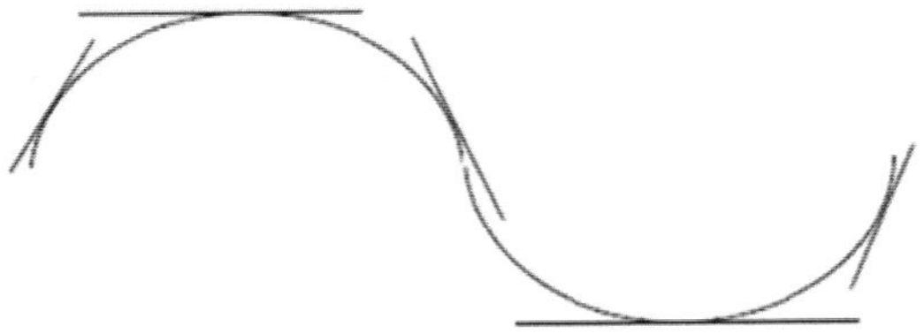

Figure 2. Gradient on the curve as the first order differential process

The result of Laplace operation is better than the gradient operation. The zero crossing that is produced by Laplace operation correspond to the slope in original curve or maximum value in gradient operation. It can be used to get the position of the edge accurately. The Laplace operation is written in equation (2) and the numerical solution is written in equations (3) and (4). Figure 3 shows the logical process on the application of differential method.

$$\nabla^2 \rho = \frac{\partial^2 \rho}{\partial x^2} + \frac{\partial^2 \rho}{\partial y^2} \quad (2)$$

$$\frac{\partial^2 \rho}{\partial^2 x} = \frac{\rho(x_{i+1}, y_j) - 2\rho(x_i, y_j) + \rho(x_{i-1}, y_j)}{(\Delta x)^2} \quad (3)$$

$$\frac{\partial^2 \rho}{\partial^2 y} = \frac{\rho(x_i, y_{j+1}) - 2\rho(x_i, y_j) + \rho(x_{i-1}, y_{j-1})}{(\Delta y)^2} \quad (4)$$

In this case, x and y are position data, i and j are the index of grid, Δx and Δy are the mesh site. Δx and Δy are extracted from the spectral analysis with FFT.

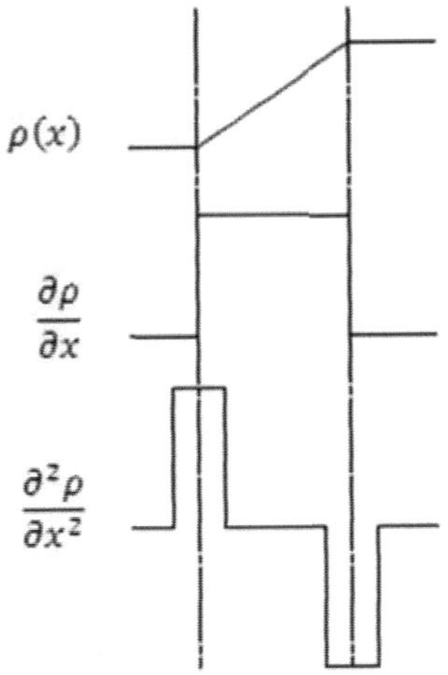

Figure 3. The illustration of Laplace operation that define the edge on 1D slope curve

3 Results and Discussion

This developed technique has been tested by experimental data that has been known and interpreted as a model. The surveyed field contains gravel, clay and sand layer [8]. It has been confirmed by seismic refraction data. The challenge is how to determine where the contacts of gravel, clay and sand are. Spectral analysis and Laplace operator has been done to find the edge of the resistive body.

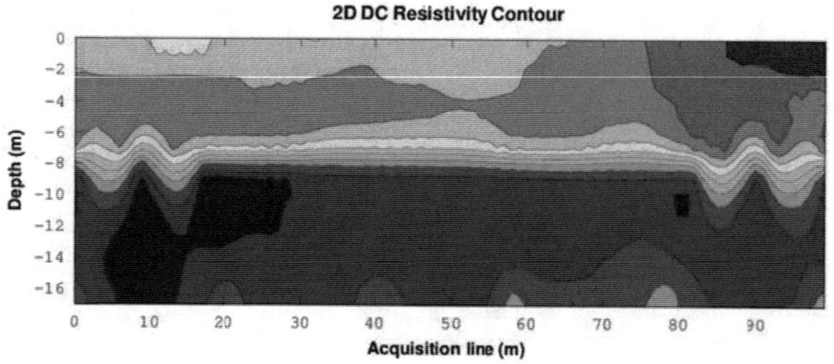

Figure 4. Resistivity contour image from Clifton survey that has gravel, clay and sand layer as a model

Figure 4 shows resistivity contours image that have high resolution grid. The grid was built by meshes of 0.2 meter. Total acquisition length is 100 m end depth is 18 m. The distance between electrodes is 2 meter. The contour appears to have two layers that show a high and low resistivity, respectively. In fact, there are three layers vertical respectively. Robust inversion method has been used to find the layers [9]. Here, the data is used to examine our technique.

The first step on the edge detection using Laplace operation and spectral analysis is to slice the contour data and get the resistivity profile. Slicing data can be digitized randomly, but fine digitization is better than coarse one. After that, the resistivity data is calculated by spectral analysis to get wave number and find the interval gridding optimum ΔxandΔy. Figure 5 shows the spectral analysis process. High wave number is interpreted as noisy signal response and it will make the Laplace operation very costly because gridding interval becomes very small. Low wave number will make the Laplace operation is very coarse and low resolution so the edge cannot be determined exactly, but numerical process is conducted faster than high wave number value. The optimum value is determined as the intersection of the linear trends at low wave number and high wave number. It is similar to the cut off frequency in filter processing [10].

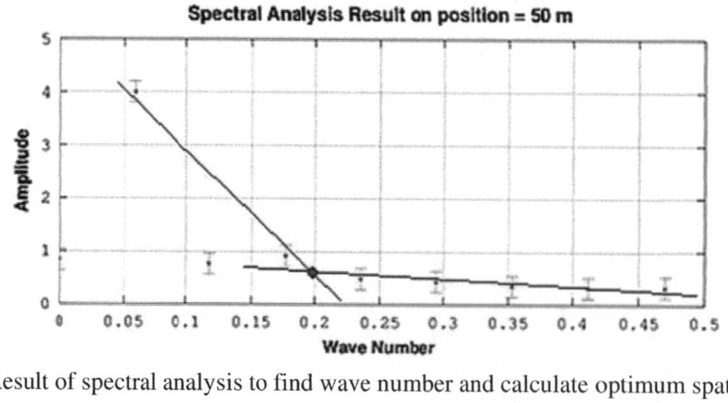

Figure 5. Result of spectral analysis to find wave number and calculate optimum spatial gridding

After the spectral analysis process, the next step is Laplace operation process. Numerical solution is very helpful in this process. The second order differential can be solved by forward and backward difference. Laplace operation process is conducted to resistivity data contour that has been gridded by new spatial gridding. It should have less data than the original one, but it does not lose the contact body information. The result of this process is the map of the new contour line that has zero value. The map shows the zero crossing made by Laplace operation and the position of the edge of the resistive body. Figure 6 shows zero value map as a position of body contact.

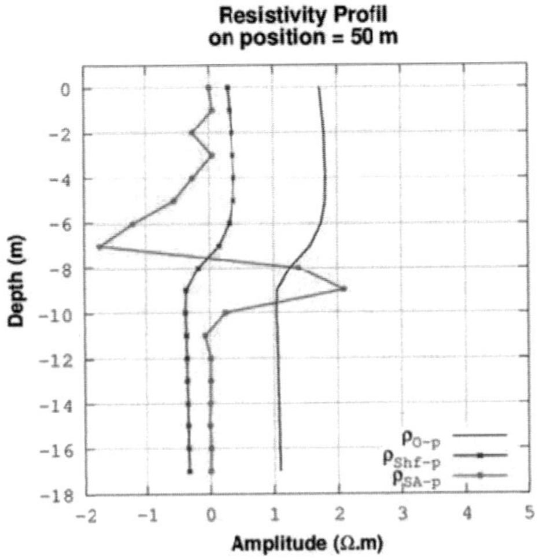

Figure 6. Resistivity profile extracted from contour on acquisition position of 50 m. The black line is original data before shifting. The blue line is data shifted across the $\rho = 0$. The red line is the result of Laplace operation.

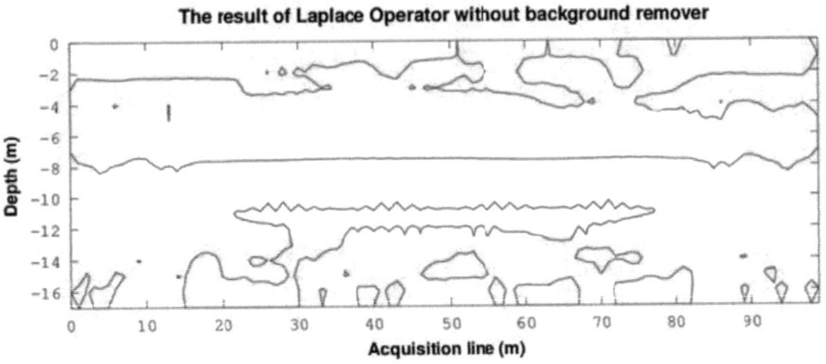

Figure 7. Zero crossing map that resulted by Laplace operation on resistivity contour to determine the edge of body contact.

In Figure 7, we can see various zero contour lines. At this paper, we don't use removal background technique to remove zero value that caused by Laplace operation in homogeneous resistive body. A homogenous resistive body can appear as noise in Laplace operation because a flat line that has no slope could lead to zero value when Laplace operation was conducted. To resolve this problem, background removal is important, but it will be discussed on another paper.

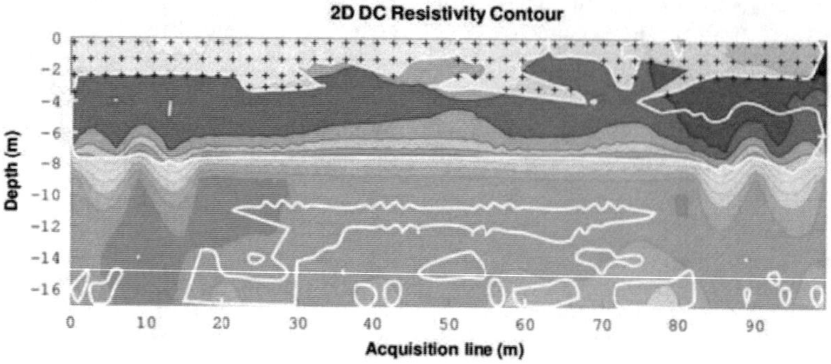

Figure 8. The result of gravel-clay-sand boundaries model overlay without background removal

Figure 8 shows model boundaries as a result of the edge detection technique on gravel-clay-sand layers. We can see the boundaries clearly, so the delineation process was successful on this model. As a note, the process has been done without background removal for noise clearance. The method can be developed to calculate the prospective natural resources exactly. Boundary detection can used to calculate the volume of prospect, so the saving of resources can be predicted accurately [11].

4 Conclusions

The combination of the spectral analysis and the Laplace operation give the boundary of resistive bodies clearly at depth 8 meter from the surface. It is demonstrated by Clifton survey data that has three layers, gravel underlying clay and sand layer. This model has been accepted as a subsurface interpretation.. Besides, the computational process to calculate the edge of body is faster because it is processed by optimal value of mesh size. In future work, this method can be used to calculate the body prospect in subsurface mapping based on 2D resistivity imaging.

References

[1] Michael V Schoor 2002 Detection of sinkholes using 2D electrical resistivity imaging. *Journal of Applied Geophysics* **50** 393-399.

[2] Acworth RI 1987 The development of crystalline basement aquifers in a tropical environment. *Quarterly Journal Engineering Geology*, **20**, 265-272.

[3] Y Srinivas et al, 2008 Resistivity studies to delineate structural features near abhishekpatti, Tirunelveli, Tamil Nadu, India. *Jounal of India Geophysics Union* **vol.12, No. 4,** pp. 157-163.

[4] Claerbout J F and Muir F 1973 Robust modeling with erratic data. *Geophysics*, **38**, 826-844.

[5] Zhonghua Ma, Dejun Liu, et al 2012 Numerical Simulation of a Multi-frequency resistivity logging-while-drilling tool using a highly accurate and adaptive higher-order finite element method. *Advance in Applied Mathematics and Mechanics* **vol. 4, no. 4,** pp. 439-453.

[6] Griffiths DH, Turnbull J and Olayinka A I 1990 Two-dimensional resistivity mapping with a computer-controlled array. *First Break*, **8**, 121-129.

[7] Loke M H 2001. Tutorial : *2-D and 3-D electrical imaging surveys*. Geo-tomo Software, Malaysia.

[8] Scott JBT, Barker RD and Peacock S 2000 Combined seismic refraction and electrical imaging. *Procs. 6th Meeting of the European Association for Environmental and Engineering Geophysics*, 3-7 Sept. 1997, Bochum, Germany, EL05

[9] Loke MH, Acworth I and Dahlin T 2003 A comparison of smooth and blocky inversion methods in 2D electrical imaging surveys. *Exploration Geophysics*, **34**, 182-187.

[10] Proakins JG and DG Monakins 1995 *Digital Signal Processing: Principles Algorithms and Applications*. Upper Sadler River : New Jersey.

[11] K Susanto and Asep Harja 2012 3D Serial Sectioning Method In Induce Polarization for Host Rock Identification On Mining Prospect. *Procs. Nasional Seminar Indonesian Geophysics Association*. PIT HAGI 37 10 - 13 September Palembang. Indonesia

IOP Conf. Series: Earth and Environmental Science **29** (2016) 012011 doi:10.1088/1755-1315/29/1/012011

3D Geothermal Modelling Using Gravity Survey on Dolok Marawa, Simalungun District, North Sumatera

A Rivandi, R Destawan, Z R Fajri, W Hidayat
Geophysical Engineering Study Program, Faculty of Mineral Technology, UPN "Veteran" Yogyakarta, Indonesia

E-mail: adekurniawangf@gmail.com

Abstract. In North Sumatera, gravity method is applied to identify the geothermal model. This method measured the earth gravitational field. This research has 160 measurement points covering 9 square kilometers. We obtained complete Bouguer anomaly values ranging 85 mGal – 130.68 mGal interpreted as a heat source of andesitic igneous rocks that are affected by the presence of Mount Bahtopu magma chamber. We interpreted the values between 40 mGal - 80 mGal as reservoir and caprock. The 3D gravity inverse modelling conducted using Gravblox, and identifying the following lithologies; Toba Pyroclastic Fall (Qjt) with density 1.92 g/cm^3, Toba Pyroclastic Flow (Qjt) with density 2.00 g/cm^3, Mount Bahtopu Andesite (Qlb) with density 2.4 g/cm^3, and 2.6 g/cm^3 which is interpreted as heat source in form of andesitic rock and Mount Bahtopu magma chamber. This heat source is estimated to be at a depth of 1.45 km to 3.78 km below the surface.

1. Introduction

Geothermal energy was formed by volcanic activity and considered as alternative energy in the replacement of fossil fuels that have limited reserves. The geothermal component consists of a heat source, reservoir rock, caprock, fluid, and structure. Dolok Marawa geothermal field located in Simalungun district, North Sumatra is characterized by the presence of hot springs and travertine deposition. Therefore, to utilize the prospect area, a proper exploration method that can describe the subsurface condition of a thermal system is required.

One of the geophysical methods that are capable of describing subsurface conditions of a geothermal system is gravity method. Gravity method can delineate subsurface conditions based on the variation of rock density by measuring gravity differences. The investigation was conducted to determine geothermal structures that exist in Marawa Dolok region, Simalungun District, North Sumatra.

2. Literature Review

The utilization of geological information for the geophysical survey in the study area is imperative. It helps to correlate the origin forming of an exploration object or target, for instance geothermal. Geologically, Dolok Marawa has the following lithologies unit from old to young in chronological order; Limestone Unit Bahbotala (Tgb), Andesite Mountain Sipapagus (Qls), Andesite Mountain Bahtopu (Qlb), Flow Pyroclastic Toba (Qat), pyroclastic fallout Toba (Qjt), Travertine (Qtr) and Alluvium (Qa). Age of Andesite lava Mount Bahtopu was 1.9 ± 0.2 Myr (Pliocene). The bedrock (basement) is crystalline and older than Middle Miocene. Base on geomorphological condition, Dolok

Marawa is divided into four morphology unit, which are in the form of steep hills, gently undulating hills, moderate undulating hills, and flat morphology. Geological structure information in the study area consists of fault Bahtopu and fault Bahbotala. The faults trend are in the direction of N 325^0 E. These fault is the cause of the presence of some geothermal manifestations in the form of hot springs and deposition of travertine [1].

3. Basic Theory

Gravity method is a geophysical method that measures the variations in the gravitational field of the earth [2]. The difference of the gravitational field occurred by mass variations in the earth's crust. The goal of this method is to associate the variations of the gravitational field to density distribution and type of rocks. The distribution of non-uniform density is caused by existing geological structures beneath the earth's surface. This method is splendid to discover the geological subsurface configuration on a broad scale based on the difference in density of each rock.

The following factors that are influenced the Gravitational value such as latitude position, the position of the sun and moon to the earth (tides), elevation, topographical condition surround area, and subsurface rock mass density variation. Gravity measurements intend to determine the value of gravity at the observation point towards a benchmark. Thus, the value of gravity observation point in the reference frame that is equal to the value of the gravitational reference point can be determined. In order to determine the value of gravity observation point one needs to do the following steps: (1) Measurements calibration; (2) Gravity meter readings conversion; (3) Feedback conversion; (4) Instrument height correction; (5) Latitude correction; (6) Earth tidal correction; (7) Drift correction.

4. Methodology

The study was carried out in the area Dolok Marawa, Simalungun District, North Sumatra Province on 6 to 30 September 2006 with coordinate 473883.442 - 483081 Easting and 342949 – 351889 Northing. We used the gravitymeter Lacoste & Romberg inc. g-802 models with 160 measurement points as shown in Figure 1. The data is processed using Ms. Excel, Pasut Software, Rockwork 99, Microdem, Gravi900, Surfer10, Oasis Montaj, and Gravblox for 3D modeling.

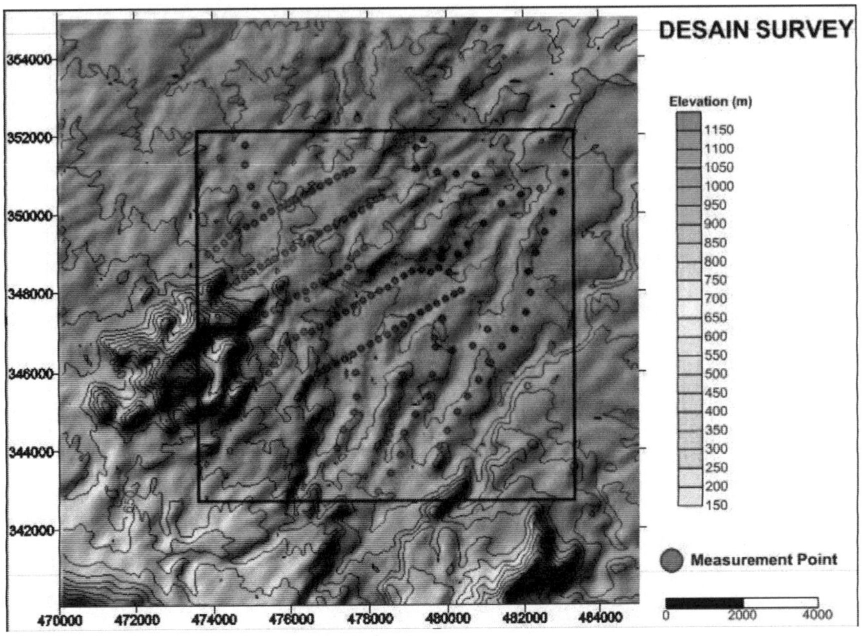

Figure 1. Design Survey

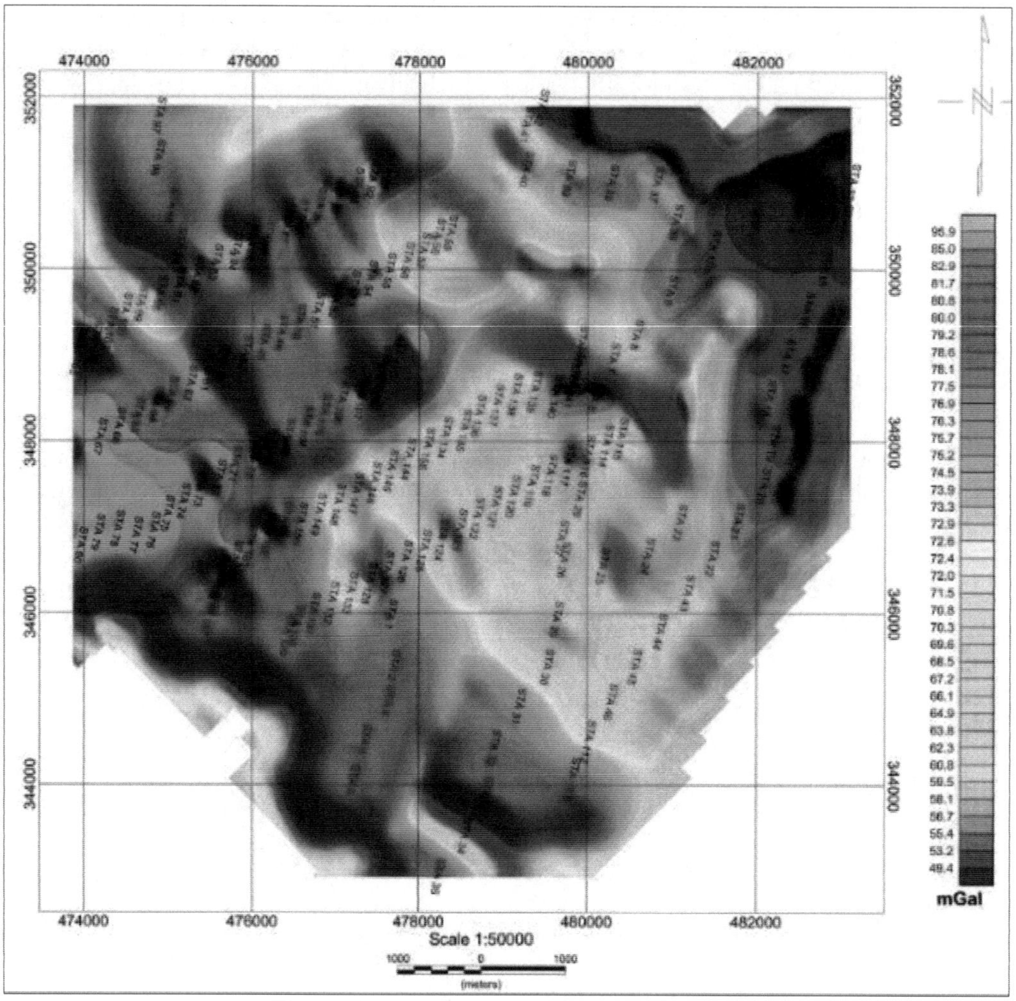

Figure 2. Complete Bouguer anomaly map

5. Results and Discussion

The gravitational variation in study area represented in the form of complete Bouguer anomaly values. These values correspond to the distribution of density in the subsurface. We interpret the geothermal component in map view (Figure 2) according to Bouguer anomaly values as the following classification.

The Heat source has high Bouguer value of 85 – 130.86 mGal as andesitic igneous rock and is influenced by Mt. Bahtopu magma chamber in western part of research area, while reservoir and caprock have low Bouguer value of 40 – 80 mGal. This variation occurs due to the density distribution of the heat source that composed of andesitic igneous rock and has higher Bouguer value than caprock and reservoir. Meanwhile, the low anomaly value is associated with the presence of fluids that filled the fracture causing a density reduction of rocks. It is hard to discriminate caprock and reservoir since it has a close range of density value. It corresponds to a gravitational theory where density is directly proportional to gravitational acceleration.

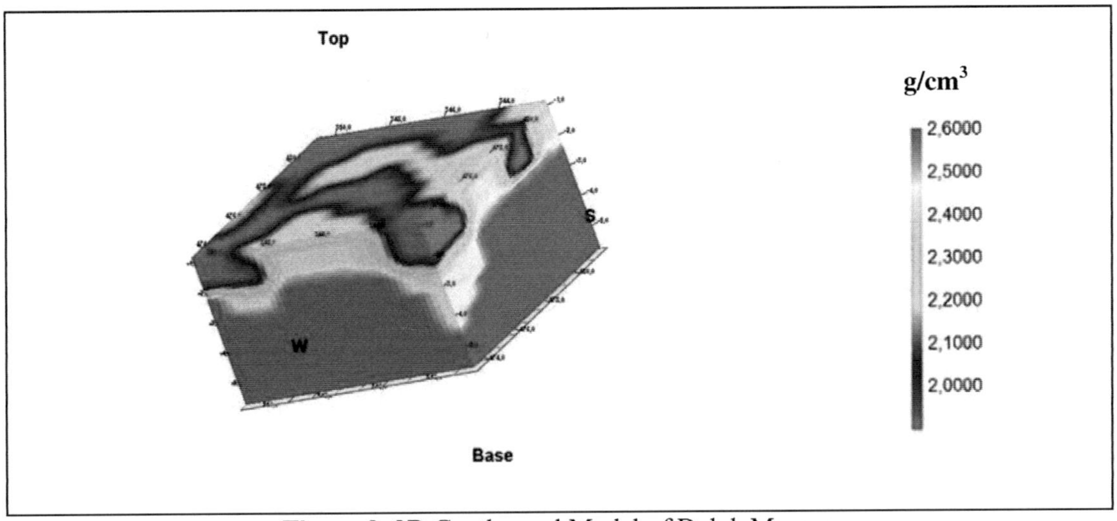

Figure 3. 3D Geothermal Model of Dolok Marawa

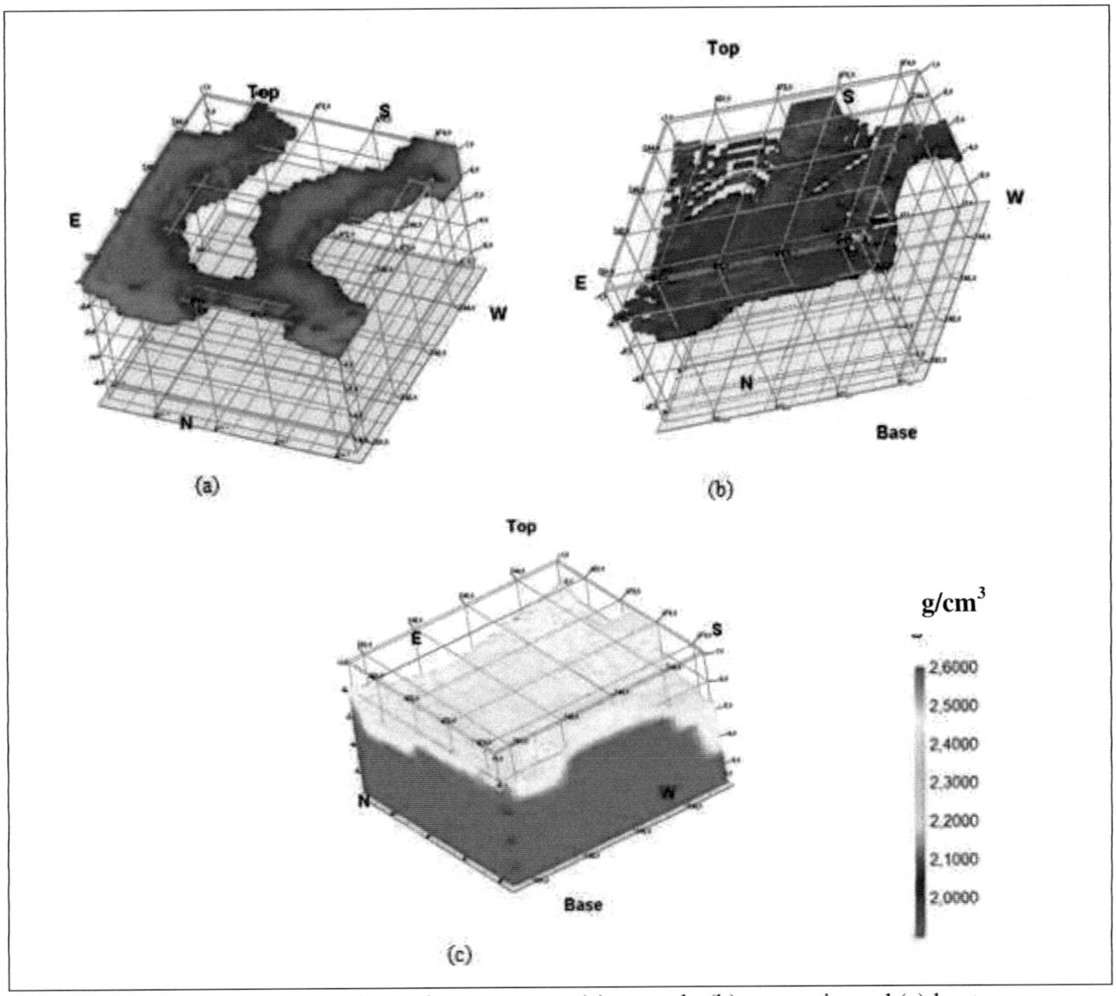

Figure 4. 3D view of geothermal components (a)caprock, (b) reservoir, and (c) heat source

Afterward, we need to analyze the position of the geothermal components and its density. It is more accurate to analyze the geothermal components in 3D model instead of 2D map view. Therefore, the inversion applied to generate the 3D model (Figure 3) and convert the Bouguer values into density.

According to 3D model, the determined components are caprock, reservoir, and heat source. In Figure 4 we focused on the certain component to have a clearer view of its propagation. According to 3D model, the heat source is located at 1.45 – 3.78 km depth represented by yellow color in 3D model with a density of 2.42 g/cm^3. Reservoir rock located at a depth of 1 – 1.99 km with a density of 2.21 g/cm^3 represented by blue color. Caprock located at a depth of 1 km with a density of 1.92 g/cm^3 and represented by magenta color. This interpretation corresponded with a density of coring data in the study area as shown in Table 1. The coring data are taken randomly to represent some of the formations in the study area.

Table 1. Density of Rock Samples

No.	Sample code	Rock	Density g/cm^3
1	7	Dacite	2.15
2	C-750	Dacite	2.15
3	19	Andesitic	2.28
4	45	Andesitic	2.44
5	R-48	Andesitic	2.28

6. Conclusion

Employing a 3D gravity model, we have successfully obtained a reasonably estimated depth of heat source, reservoir, and caprock of geothermal system instead of 2D map views in Dolok Marawa Region. The results are confirmed by coring and correspond to the modeling result. Generally, the geothermal component in Dolok Marawa are composed mainly of three types of density parameters that are supported by core data and can be explained as follows; (1) The heat source has density of 2.42 g/cm^3 and Bouguer values of 85 – 130.68 mGal at 1.45 – 3.78 km depth. (2) The reservoir and caprock cannot be discriminated in the complete Bouguer anomaly map with values of 40 – 80 mGal but are obviously separated in a 3D model. (3) Caprock has a density of 1.92 g/cm^3 at up to 1 km depth, and reservoir rock has a density of 2.21 g/cm^3 at 1 – 1.99 km depth.

References

[1] Lim, D., Setiadarma, Sundhoro, H., dan Sulaeman, B., 2006. Penyelidikan Geologi Dan Geokimia Di Daerah Panas Bumi Dolok Marawa, Kabupaten Simalungun – Sumut. Bandung: Pusat Sumber Daya Geologi

[2] Telford, W.M., Geldart, L.P., and Sheriff, R.E., 1990. Applied Geophysics second edition. Cambridge: Cambridge Univ. Press.

IOP Conf. Series: Earth and Environmental Science **29** (2016) 012012 doi:10.1088/1755-1315/29/1/012012

Mapping the fissure potential zones based on microtremor measurement in Denpasar City, Bali

U N Prabowo[1*], Marjiyono[2] and Sismanto[3]

[1] Postgraduate of Physics Department, Gajah Mada University, Sekip Utara Bulaksumur, Yogyakarta, Indonesia
[2] Geological Survey Institute, Geology Agency, Energy and Mineral Resource of Indonesian Republic, Diponegoro Street 57, Bandung, Indonesia
[3] Department of Physics, Gajah Mada University, Sekip Utara Bulaksumur, Yogyakarta, Indonesia

email: urip.nurwijayanto@gmail.com

Abstract. Denpasar City and its vicinity considered as the areas with excessive ground water exploitation and high earthquake intensity. These conditions will cause these area potential with land subsidence which is triggering ground fissures. This research aims are to mapping the fissures potential areas based on microtremor measurement in Denpasar City and its vicinity. Ground fissures will happen if the land subsidence occurs in the areas which have different bedrock height beneath its sedimentary layer. The height of bedrock is determined by reducing surface elevation with the sedimentary layer thickness. This sedimentary layer thickness obtained from microtremor measurement using HVSR method, and Shear wave velocity (V_S) obtained from microtremor array measurement which is analyzed by Spatial Auto Correlation (SPAC) method. The result from HVSR method as well as Peak Ground Acceleration (PGA) value are then analyzed to get ground shear strain value, which is the soil surface strain and its effect when earthquake occurs. Based on the bedrock map, it can be estimated that the bedrock layer forms structure in the southern part of the research areas and the potential fissuring area due to the massive ground water exploitation is in the west Denpasar Subdistrict. In addition, based on the bedrock map and ground shear strain value which combined with Simple Additive Weight (SAW) method, there are two areas having fissuring potential, i.e west and south Denpasar Subdistricts.

1. Introduction

Denpasar city and its vicinity were considered as the main tourism destination in Bali with rapid development and high water consumption. Almost 30% water supply in Denpasar city provided by government water department from ground water exploitation [1]. Beside that, hotel and tourism business had their own ground water exploration well to fulfil their water needs, this area also have high earthquake intensity, about VI-VII MMI [2]. This condition can cause sedimentary layer compaction at these areas and land subsidence will occurs. Land subsidence is the ground elevation lowering phenomenom caused by soil compaction related to ground water depletion or earthquake. Ground water loss that fills the space between soil grains can cause sedimentary layer compaction and land subsidence will occurs. Earthquake can also caused by the land subsidence when earthquake

* To whom any correspondence should be addressed.

shake the saturated or partially saturated soil then increase pore water pressure and decrease effective soil stress causing water came out from the soil. This called liquifaction phenomenom. Land subsidence that occurs in small area which has different bedrock height beneath sedimentary layer (high slope of bedrock) will increase the horizontal strain of sedimentary layer and triggering ground separations or fissures [3].

The bedrock height (Z_B) as the main parameter that cause fissure when land subsidence occurred. It could be investigated using drilling borehole but it is expensive and slow. Microtremor measurement as one of the geophysical techniques can be used as an alternative that allow investigation in wide area and reduced both time and cost of investigations. Microtremor is all ground vibration not due to earthquake or explosion [4]. Source of microtremor originated from daily human activities such as factory machinery movement, motor, cars, people walking; and natural phenomenon such as rain, wind, ocean waves, and variation in atmospheric pressure [5]. Shelley et.al [3] showed that sharp step of bedrock derived from microtremor related to fissures that occured in Mexico city and the measurement can used to predict the area of next fissure will occur. Microtremor measurement is used to map sedimentary layer thickness, which is the layer on bedrock, and it measurement can be very useful and accurate in mapping small variation in shape and geological structure of bedrock [6]. Sedimentary layer thickness obtained using microtremor measurement using equation

$$H = \frac{V_S}{4 f_0} \tag{1}$$

where H is sedimentary thickness, f_0 is the dominant frequency of sedimentary layer obtained from single station microtremor measurement, and V_S is the shear wave velocity of sedimentary layer obtained from array microtremor measurement.

Single station microtremor measured 3 components of microtremor on sedimentary layer and was analysed using Horizontal-to-Vertical spectral ratio (HVSR) method. This method was proposed by Nakamura [7]. Dominant frequency of sedimentary layer known from H/V curve that obtained by dividing the average spectra of horizontal components by vertical component of microtremor signal. This calculation can eliminate the influence of source effect and also can determined soil amplification factor (A) when earthquake occur since H/V curve exhibits a good agreement with H/V curve from earthquake recording. Further analysis using dominant frequency and amplification, Nakamura [8] using these values and peak ground acceleration or PGA (α) of bedrock layer [9] to determine ground shear strain value (γ), which is the soil surface strain value, calculated by equation

$$\gamma = \frac{A^2}{f_0} \times 10^{-6} \times \alpha \tag{2}$$

Fissures will occur in area which has ground shear strain values about 10^{-4}-10^{-2} [8]. Daryono [10] showed that fissures in Yogyakarta caused by earthquake in 2006 occured in an area which has high ground shear strain value.

Array microtremor measured vertical component of microtremor to estimate a subsurface layered-earth structure and shear wave velocity (V_S) profiles on the assumption that microtremor consist of predominately of Rayleigh waves [5]. Spatial autocorrelation (SPAC) method is widely used for array microtremor data collection, processing and analysis. This method gives good result but need less recording station and shorter dimension arrays measurement compared with the other method. The spatial autocorrelation (SPAC) method was first introduced by Aki [4] that uses the spatial and time stationaries of microtremor to calculate average spatial coherency ($\bar{c}$) over different stations separations that has the shape of zero order Bessel function (J_0) show by equation

$$\bar{c}(f) = J_0\left(\frac{2\pi f r}{V_S(f)}\right) \tag{3}$$

where *r* is seismometer separation distance. The coherency curve is fitted to Bessel function by iterative inverse modelling to obtained shear wave velocity of sedimentary layer [11].

In this work, we developed microzonation of fissure hazard when earthquake and excessive ground water exploration happens. Microtremor measurements was derived to mapped the bedrock height as the main parameter to determine fissure potential zone when excessive ground water exploitation happen and using simple additive weight analysis. We combine the map of bedrock height and ground shear strain value as parameters to determine fissure potential zone when the earthquake happen. SAW is a multicriteria analysis method that combines some data parameter (attributes) into one decision output. This decision output calculated by multiply the alternative of each attribute with the attribute value, then sum the multiply value for all attribute [12].

2. Geological Setting

Geological setting in the middle and south part of the area are included in Buyan-Bratan and Batur volcanic formation (Qpbb) comprises Holocene age that consist of tuff and volcanic mudflow. South part of the area is included in alluvium formation comprises Holocene age that consist of sandstone, clay stone, shale stone, and mix of rivers, seas, and lake precipitations [13].

3. Single Station Microtremor

Single station microtremor measurements of 176 sites (figure 1.a) used Mark L4-3D seismometer were carried out over Denpasar city and its vicinity from May - June 2014 and October - November 2014. The HVSR method has been applied on data that were acquired, selected, processed according to the recommendations of Sesame European Project and computed using geopsy free software. Before the spectra computation, the window length for each data was set to a period of 15 seconds in order to capture low frequency of H/V peak curve at a sampling rate of 100 samples per second and 5% cosine taper value. Geopsy software were used to compute the mean and linear trend of all windows, then transformed into Fourier spectra and smoothed by Konno and Ohmachi one with b=15.

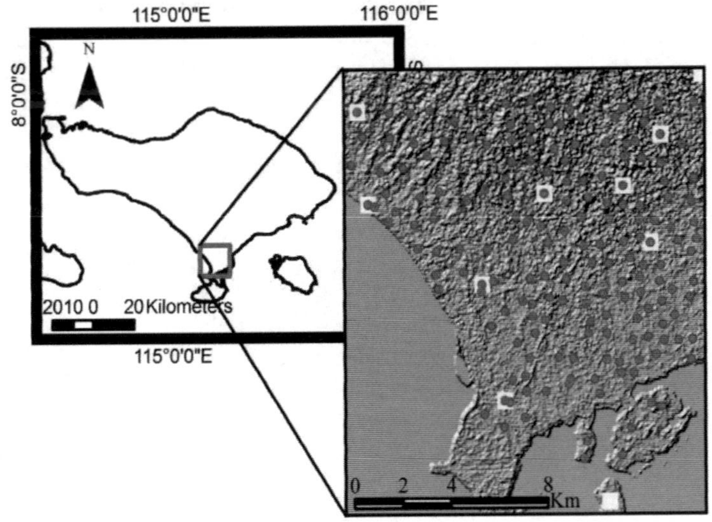

(a)

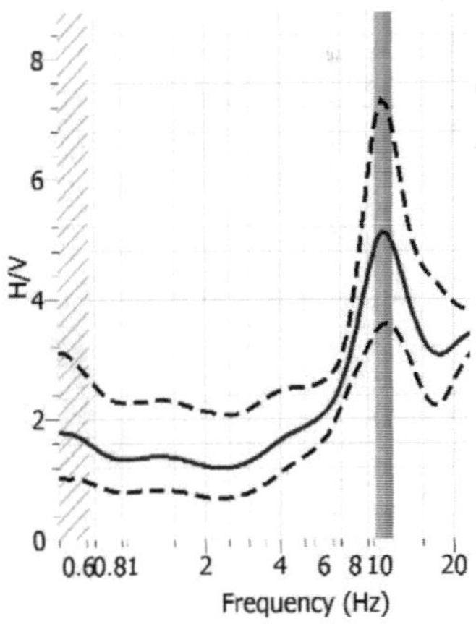

(b)

Figure 1. (a) Distribution of measurement points, the red dots are single station microtremor points and yellow squares are array micretremor points. ASTER GDEM map from USGS, (b) H/V curve of point A001.

Most of the H/V curve results show very clear single peak (figure 1.b) with amplitudes that related to amplification value about 1.73-11.80 (figure 2.a). This amplification value could be correlated with a quite clear impedance contrast between bedrock and overlying sedimentary layer. A predominant peak at dominant frequency about 1.04–14.90 Hz appears at research area. Distribution of dominant frequency (figure 2.b) present high dominant frequency value on northern part and small dominant frequency value on southern part. It represents that sedimentary thickness become thicker from northern to southern part since the dominant frequency have negative correlation with sedimentary thickness based on equation 1.

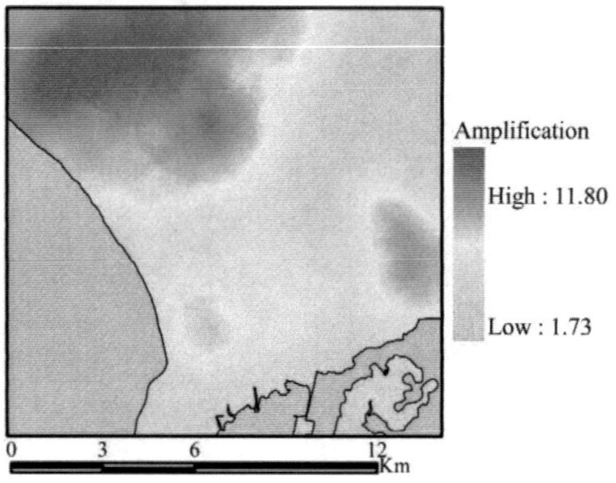

(a)

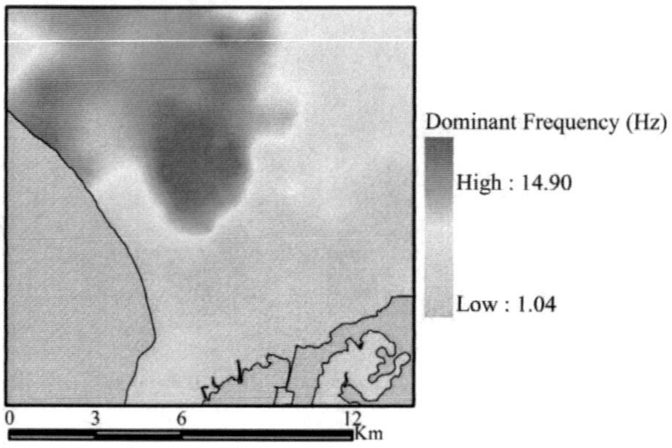

(b)

Figure 2. (a) Amplification map, (b) Dominant frequency map.

4. Ground Shear Strain

The ground shear strain calculated using equation 2 and figure 3 show the distribution of its value. The parameter that used for PGA calculation are the hipocenter and magnitude of 6.41 *Mw* earthquake that occoured in July 14, 1974. The result of PGA calculation in this research area are about 49.03-65.13 gal. The distribution show that based on Isihara's classification [8], almost all the area have high fissure vulnerability because it has ground shear strain value about 10^{-4}-10^{-2} (yellow colour). The West Denpasar subdistrict (green colour) have low ground shear strain value because this area has low amplification value (low contrast of sedimentary layer and bedrock) and high dominant frequency value that represent thin sedimentary layer.

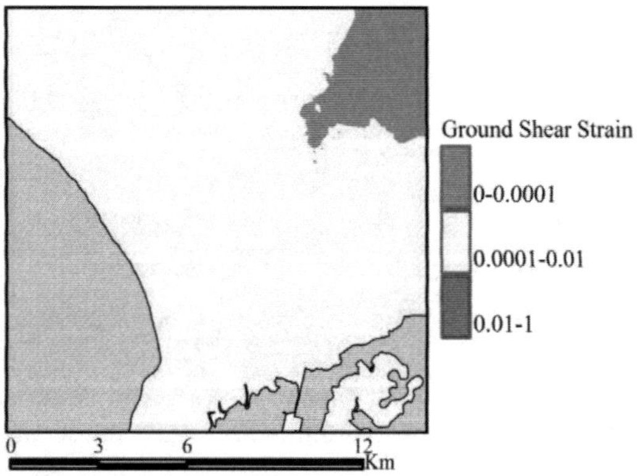

Figure 3. Ground shear strain distribution map.

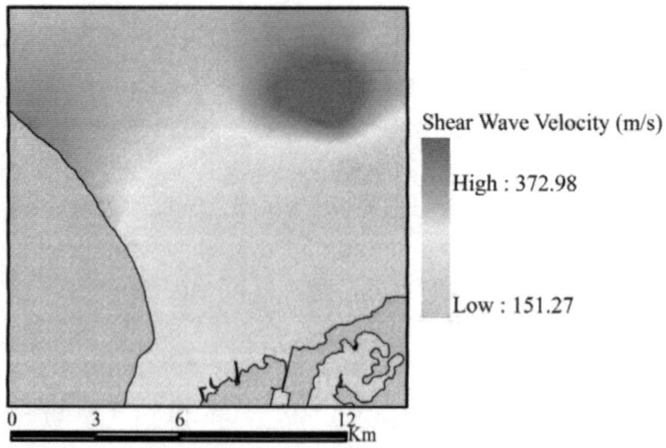

Figure 4. Sedimentary layer shear wave velocity map.

5. Array Microtremor

Array microtremor measurements of 13 sites (figure 1) used 4 OYO seismometer 1134 model were carried out over Denpasar city and it is vicinity from May - June 2014 and October - November 2014. The measurement used triangular array configuration, with seismometer distance (*r*) 2 m, 5 m, 7 m, 10 m, and 20 m for each sites. The duration of the measurement are 20-45 minutes depend on the seismometer distance and its sampling rate are 100 samples per second. The SPAC method has been applied on the acquired data, then processed and computed using geopsy, spac2disp and dinver which are included in geopsy free software package. Shear wave velocity of sedimentary layer resulted from SPAC method (figure 5) then interoplated to obtained sedimentary layer shear wave velocity map. The depth value haven't used for the further analysis because the inversion of homogen sedimentary layer assumption would resulted far estimation from real depth but the shear wave velocity was well estimated compared to borehole measurement [11]. Sedimentary layer shear wave velocity map (figure 4) showed that southern part of the area has lower shear waves velocity than northern part of the area. This condition was caused southern part are alluvium sediment that is weaker than northern part which are Buyan-Bratan volcanic sediment.

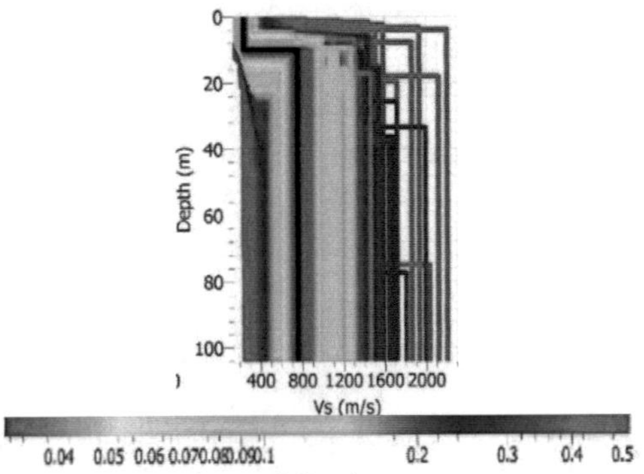

Figure 5. sedimentary layer shear wave velocity of point A01 resulted from SPAC method

6. Bedrock Morphology

The bedrock morphology of the area (figure 6.a) estimated by reducing the surface elevation from USGS Aster GDEM map with the distribution of sedimentary thickness calculated using equation 1. The bedrock morphology results has compared with drilling result [14] (figure 6.b, figure 6.c, figure 6.d). A-A' and C-C' profiles show that the bedrock morphology has good correlation with the drilling result. B-B' profile show bad correlation of bedrock morphology with the drilling result. It happen because the area around B-B' profiles has large dominant frequency value about 1.90-4.35 Hz that represents high variation of sedimentary thickness at this area. A detailed measurement is needed in this area. Figure 6.a shows that southern part of the area has low bedrock height (green colour), located in south part of the area between Mengwi subdistrict, West Denpasar subdistrict and South Denpasar subdistrict. This bedrock height contrast with the other area around it and we predict that the bedrock formed a basin structure. Soebowo et. al [14] named this basin structure as south basin of Bali and this basin predicted as the result of faulted graben in south Bali. We estimated the fault orientation that resulted the graben by the dashed blue line showed in figure 6.a.

7. Estimation of Fissure Potential Zone

Slope of bedrock map (figure 7) show the high fissure potential zone when excessive ground water exploration happen is West Denpasar Subdistrict which has slope value >45^0 (red area) and located at basin structure. Medium fissure potential zones are West and South Denpasar subdistrict, a part of East Denpasar subdistrict and Mengwi subdistrict. High and medium fissure potential area located in pedestrian area included in Buyan-Bratan formation and small area of alluvium formation. Almost all the area has fissure potential when excessive ground water exploitation happen because from Bali

IOP Conf. Series: Earth and Environmental Science **29** (2016) 012012 doi:10.1088/1755-1315/29/1/012012

hydrology map [15] was known that bedrock height are lower than ground water height except the northern part of this area which has ground water heigh 25-35 m.

Fissure potential zone when earthquake happen (figure 8) has been estimated based on the bedrock slope value and ground shear strain value which both of them combined using simple additive weight method. High fissures potential zones when earthquake happen are in the West Denpasar subdistrict, South Denpasar subdistrict and small east part of Kuta subdistrict. Fissures can occur at this area because from hydrology map of Bali [15] shows that the surface elevation gap with ground water height is less than 5 m. This condition can trigger liquefaction when earthquake happen and causing fissure because it has high ground shear strain value and high slope of bedrock.

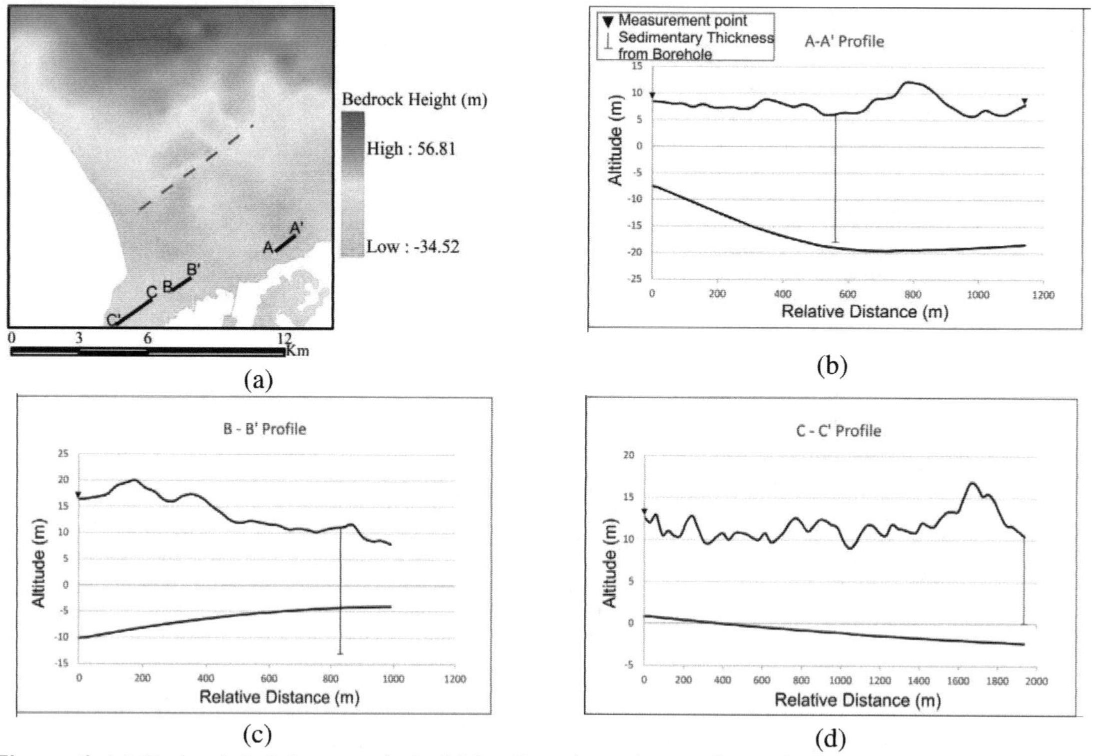

Figure 6. (a) Bedrock height map, dashed blue line show the predicted fault, (b) A-A' (c) B-B' (d) C-C' cross sections showing the subsurface structure profile.

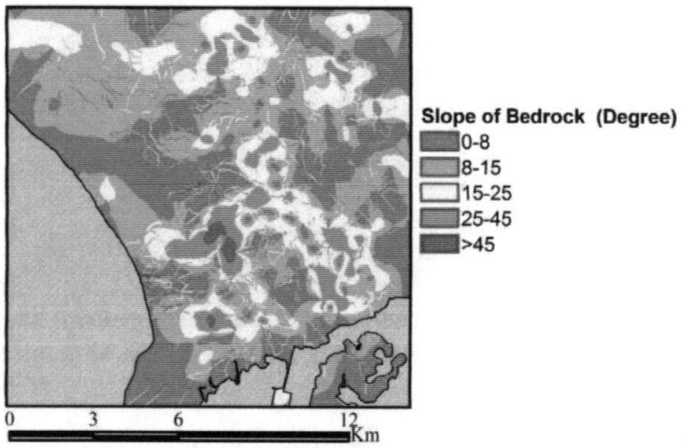

Figure 7. Slope of bedrock height map.

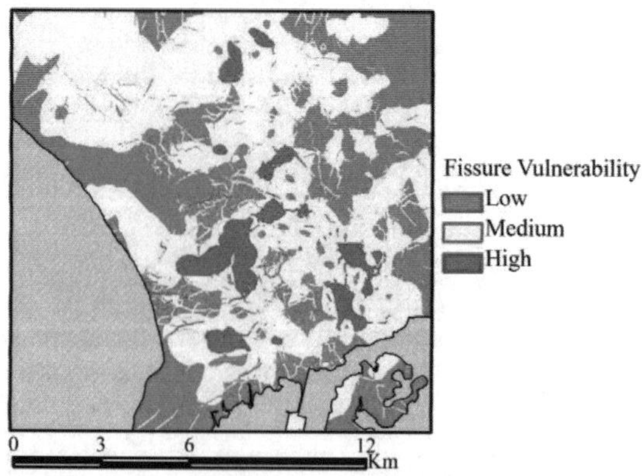

Figure 8. Fissure potential map when earthquake happen

8. Conclusion

In the present research, microtremor analysis have been carried and the fissure potential zone of the research area have been determined. Our result confirm the suitability of microtremor analysis as geophysical method to explore the geological setting especially in sedimentary layer. This analysis can also well estimate the bedrock morphology underlying the sedimentary layer. In addition, we has estimated the fissure potential zone that can be used for fissure and seismic hazard assesement. As the results, there are two areas having fissures potential, i.e West and South Denpasar subdistricts.

Acknowledgement

Pusat Survey Geologi (PSG) for the data that had been used in this research.

References

[1] http://bankdata.denpasarkota.go.id, accesed on 12 November 2014.

[2] Supartoyo, G. Suantika and Djadja, 2009, *Peta Kawasan Rawan Bencana Gempabumi Pulai Bali*, Pusat Vulkanologi dan Mitigasi Bencana Geologi, Bandung.

[3] Shelley, E.O., Javier L.S., Gabriel A., and Edgar M.S., 2012, Microtremor Measurements To Identify Zones Of Potential Fissuring In The Basin Of Mexico, *J. Geofisica Internacional* **51** 2, 143-156.

[4] Aki, K., 1957, Space and Time Spectra of Stationary Stochastic Waves, with Special Reference to Microtremors, *Earthquake Research Institute Japan.*

[5] Okada, H., 2003, The Microtremor Survey Method (Geophysical Monograph Series Number 12), *Society of Exploration Geophysicists*, Amerika.

[6] Petermans, T., X. Devleeschouwer, F. Pouriel and P. Rosset, 2006, Mapping the Local Seismic Hazard in the Urban Area of Brussels, Belgium, *IAEG*, 424.

[7] Nakamura, Y, 1989, A Method for Dynamic Characteristic Estimation of Subsurface using Microtremor on the Ground Surface, *QR Railway Technical Research Institute* **30** 1 25-33.

[8] Nakamura, Y, 1997, Seismic Vulnerability Indices for Ground And Structures Using Microtremor, *World Congress on Railway Research, Florence* Nov. 1997.

[9] Fukhusima, Y., and Tanaka, T., 1990, A New Attenuation Relation for Peak Horizontal Acceleration of Strong Earthquake Ground Motion in Japan, *Bull of the seismological society of America. Soc. Am.* **80** 757-783.

[10] Daryono, 2011, Seismic Vulnerability Index Based on Microtremor in Various Landform Units at Bantul Graben Zone, Yogyakarta Special Province, *Disertation*, Postgraduated Program of Geography Faculty, UGM, Yogyakarta.

[11] Bettig, B., P.Y. Bard, F. Scherbaum, J. Riepl, F. Cotton, C. Cornou and D. Hatzfeld, 2001, Analysis of Dense Array Noise Measurements Using The Modified Spatial Auto-Correlation Method (SPAC): Application to The Genoble Area, *Bulletino Di Geofisica Teorica Ed Applicata* **42** 3-4 281-304.

[12] Setiawan, J.J.H., 2009, Mikrozoonasi Seismisitas Daerah Yogyakarta dan Sekitarnya, *Thesis*, Magister Program of Geodesy and Geomatika Technique ITB, Bandung.

[13] Hadiwidjojo, M.M., H. Samodra and T.C Amin, 1998, *Peta Geologi Lembar Bali, Nusatenggara*, Direktorat Jendeal Geologi dan Sumberdaya Mineral, Departemen Pertambangan dan Energi.

[14] Soebowo, E., D. Sarah, Y. Kumoro, and N.A Satriyo, 2011, Identifikasi Zona Penurunan Tanah Akibat Likuifaksi di Daerah Serangan-Tanjung Benoa, Bali Selatan, *Prosiding Pemaparan Hasil Penelitian Puslit Geoteknologi-LIPI 2011*, 193-198.

[15] Sudadi, H. Setiadi, B.R. Denny, Sarief, S. Ruchijat and S. Hadi, 1986, *Hydrogeological Map of Indonesia*, Directorate Geological and Mining Area Enviromental, Geological Agency of Indonesia, Bandung.

IOP Conf. Series: Earth and Environmental Science **29** (2016) 012013 doi:10.1088/1755-1315/29/1/012013

Graben Structure Identification Using Gravity Method

Lilik Hasanah[1], Ahmad Aminudin[1], Nanang D. Ardi[1], Agus. S. Utomo[1], Heru Yuwono[1], Kamtono[1], Dadan. D. Wardhana[2], Karit L.Gaol[2] and Mimin Iryanti[1]

[1]Physics Departement, Universitas Pendidikan Indonesia,
[2]Research Centre for Geotechnology LIPI

Email: lilikhasanah@upi.edu

Abstract. Graben (trench) is a natural expanse that is lower in altitude compared to its surrounding which is caused by normal faults shift. Changes in rock density can be identified to obtain the subsurface rock structure. Gravity method is a basic method yet very effective in determining subsurface rock structure. Identification of graben structure is the main focus of this research in order to identify the natural resources which may be available under the ground. Research work was performed in various locations in Bogor and according to our analysis using 2D Talwani model, the average density at the surface is 2.5 gram/cm^3. 2D modelling results show a fault structure at rocks with relative direction West-East. The fault is forming an extension block faulting which makes the area a graben. In general, the cross-section profile of the model indicates rock layer structure made from limestone rocks (2.75 gr/cm3), sands, flakes, limestone (2.5 gr/cm^3), volcanic sediment layer sandstone (2.3 gr/cm^3), and clay layer and similar (2.00 – 2.10 gr/cm^3).

1. Introduction

Tectonic plate activities have been affecting changes in geological structure physically in earth's crust and it appears in the West Java landscape. Tectonic force does continuously press, pull, bend, and fracture the rocks in lithosphere [1]. Subduction zone in the south of Java that is continuing to Sumatera contributed to tectonic compression that forms Meratus, Sunda, and Java Pattern. Bogor Basin, which in Middle Eocene - Oligocene was a front basin of magmatic arc, changed its structure into rear basin of magmatic arc in Early Miocene – Pliocene [2], and it formed a lot of thrust faults and creases.

Graben (trench) is a natural expanse that is lower in altitude compared to its surrounding which is caused by normal fault shift [3] Changes in rock density can be identified to obtain the subsurface rock structure. Gravity method is a basic yet very effective tool in determining subsurface rock structure. Some researchers identified graben with gravity method yet [4, 5]. Identification of graben structure is the main focus of this research in order to identify the natural resources, which may be available under the ground.

Uneven density distribution under the earth surface can be caused by the available geological structure. Contribution of the structure to the difference in gravity acceleration on earth surface is very low, compared to its absolute value. The difference is also contributed by observation's point position in earth surface, since earth is not a perfect sphere and earth's relief varies.

Gravity method is an investigation method in geophysics that is based on the variety of gravity acceleration on earth's surface. This method is used to identify and describe rocks lithology geological structure and subsurface structure depending on varied earth's gravity which is caused by difference in various rocks' density.

This research uses gravity data analysis to determine subsurface structure of fault patterns which focus in a slicing model. This is done because geophysics geology survey has not been done before in the area, so there is very limited information about the area. The slicing model is very unique to be observed because it is expected to have a normal fault block structure pattern as expanded fault. Generally, the area is formed by volcanic rocks, sediment rocks, and intrusive that can form a hydrocarbon reservoir. In research area, there is a high contrast of rock density and complex fault structure, which is predicted to be a hydrocarbon manifestation [6].

One technique that will help in interpreting the depth of the source of gravity anomaly is upward continuation that is supported by spectral analysis information in the slicing which will be tested. The purpose of the analysis is to give effective information about geological structure that will be analyzed. Thus, it is our hope that this method can be used to help collecting information about subsurface geological structure in Bogor Basin, West Java.

2. Methods

Research Centre for Geotechnology LIPI carried out data acquisition using point closed looping method. By applying some necessary corrections on the collected data, a complete Bouguer anomaly contour map was succeeded in showing some anomalies in the field (see Figure 1). The map was created by using 169 data points data which was symbolized by '+' sign on the map.

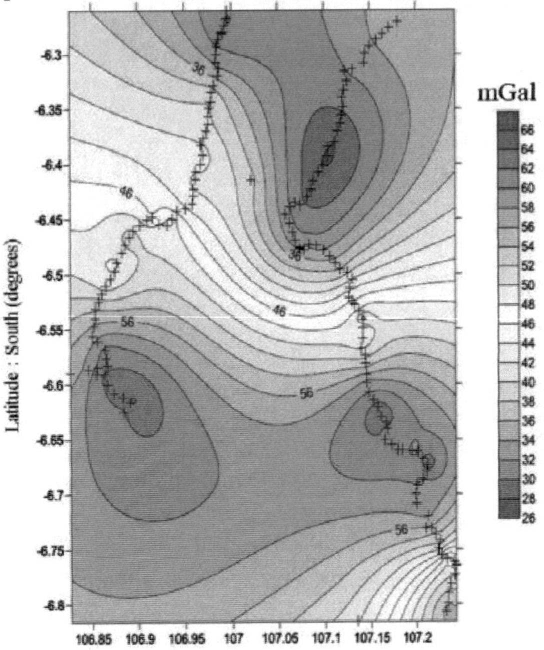

Figure 1. Complete Bouguer anomaly map

To separate noise, residual, and regional information, spectral analytic method is used in this research (see Figure 3 and Figure 4). The spectral analysis used is upward continuation method that converts potential field data measured on a certain surface level into values, which reflect data as if measured at higher surface level (see Figure 2). Upward continuation will show anomalies that are caused by shallow sources. To simplify modelling, GMSys 2D software that is based on 2D Talwani method will be used using interactive forward modelling. This modeling is performed by predicting the geometrical form of subsurface that will be correlated with geological structure.

3. Result and Discussion

This Complete Bouguer Anomaly Map has anomaly value in a range between 26 until 66 mGal, with general pattern in the form of closed and open basin with varied value. This distribution map describes shallow and deep subsurface structure. Differences of Bouguer anomaly value are marked with colour gradation, where each colour represents certain anomaly value.

Regional anomaly is associated with general geological condition of the area, characterized by low frequency anomaly. Bouguer anomaly can also be considered as a resultant/accumulation of subsurface rocks structure's influences which consist of anomalies caused by shallow, medium, and deep geological structure, including adjacent and mutually interacting structures that cause overlap anomaly as shown on the map.

The research area is an area of interest because it belongs to high anomaly group. The low anomaly is probably caused by the presence of faults that causes the area to be depressed and lower than its surrounding. The high anomaly, on the other hand, is caused by the availability of solid/compact material in the area.

In this research, contour map produced with upward continuation is not significant compared to Complete Bouguer anomaly map (ABL). This happens because the contour on ABL map is qualitatively smooth. Fixed and relatively far distance between measurement points (750 m effectively) and using only one slicing section model may contribute to the effect, which also contribute to a linear distribution in map-gridding processing phase (statistical computation).

Spectral analysis results show that the source of regional anomaly is at 1769.5-meter depth, residual anomaly source at 132.29-meter depth, and noise source at 17.7-meter depth (see Figure 2). Based on the result, limit of depth value of residual-regional anomaly is between 1558.508 and 2805.314 m, which show a high anomaly pattern caused by the presence of normal fault block structure pattern. The result from spectral analysis and upward continuation information are needed in objectively subsurface interpretation AA' slicing model (see figure 5). From the result of separation filtering, it can be concluded that the higher the continuation then the deeper the regional anomaly and the more shallow the residual one.

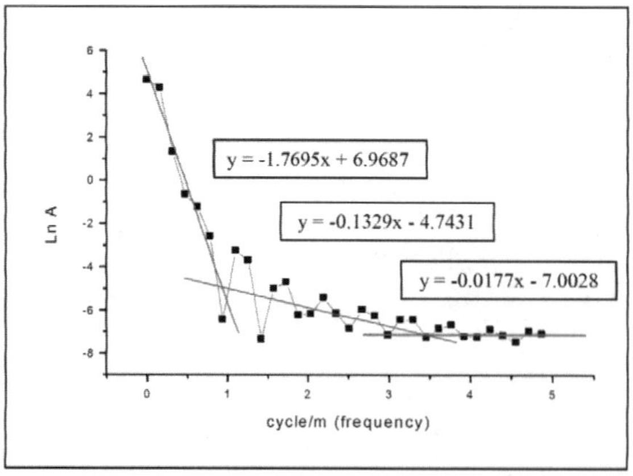

Figure 2. Spectral analysis from gravity data

There is a coherent relation between the analysis results of gravity data using upward continuation with spectral as control data information in assisting the interpretation of gravity subsurface geological structure, considering that both analytical methods are subsurface potential field data analysis application [7].

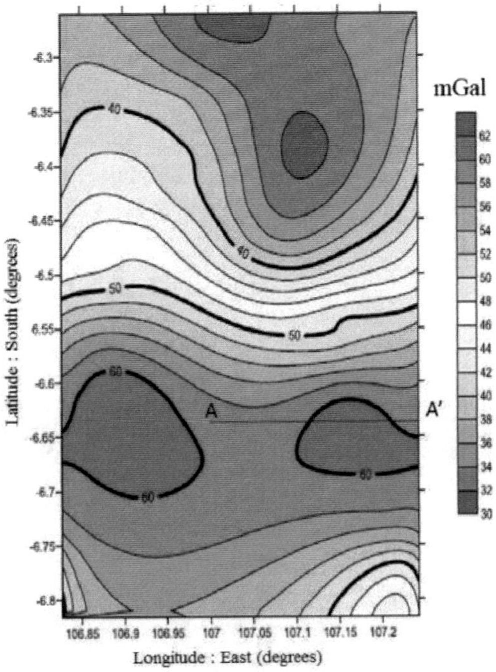

Figure 3. Regional anomaly map with upward continuation

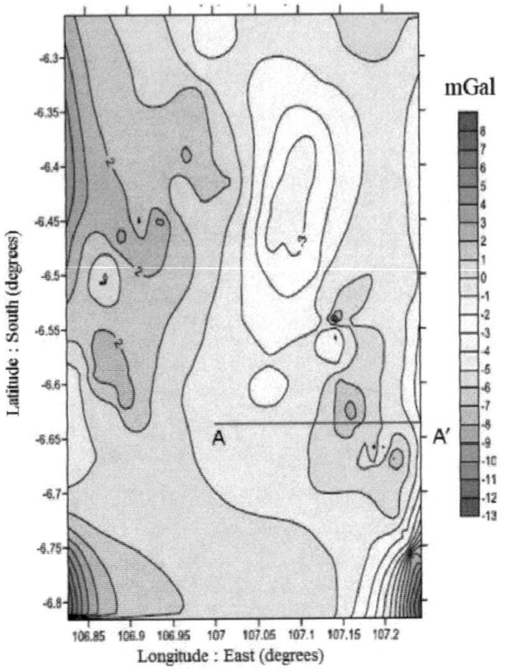

Figure 4. Residual anomaly map with upward continuation

Positive anomaly that form heights and closed basin is predicted as a result of interaction between horst-graben caused by normal fault (rifting) in West-East direction. Rocks layer from simulation in this trajectory is interpreted against average of rock density in research area, which is 2.507647059 gr/cm^3 ≈ 2.5 gr/cm^3. Based on the result, basement rock (red *body layer*), predicted to be sedimented and then covered by sediment from Quaternary volcanic, which has density between 2,14–2.52 gr/cm^3. This might be true if the type of the rocks is sand rock.

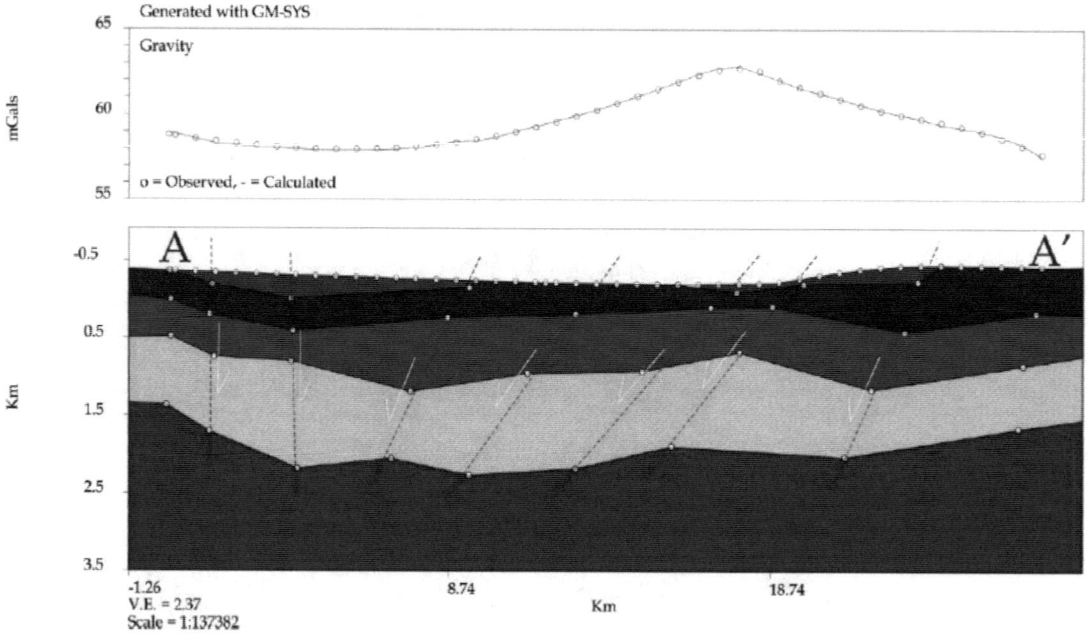

Figure 5. 2D slicing section model of research area

4. Conclusions

The source of regional anomaly according to continuation result and spectral analysis information is at depth range between 1558.508 until 2805.314 meter. The model's section cut as shown in Figure 4 has rock layer structure that consists of limestone (2.75 gr/cm^3), sand, flakes, limestone (2.5 gr/cm^3), volcanic sediment layer and sandstone (2.3 gr/cm^3), and clay and similar (2.00 – 2.10 gr/cm^3).

References

[1] Y. Krien and L. Fleitout, 2008, Gravity above subduction zones and forces controlling plate motions, *Journal Of Geophysical Research*, Vol. **113**, B09407, Doi:10.1029/2007jb005270.

[2] Aswan, Emmy Suparka, Sonia Rijani, Dessy Sundari and Emma Yan Patriani, 2008, Asymmetrical condition of the Bogor Basin (West Java, Indonesia) during the Middle Miocene to Pliocene based on Taphonomic study of Shellbed and its sequence architecture, *Bulletin of the Geological Survey of Japan*, vol.**59** (7/8), p. 319-325.

[3] Şenol Özyalın, Oya Pamukçu, Tolga Gönenç, Ayça Yurdakul and Hasan Sözbilir, 2012, Application of boundary analysis and modeling methods on Bouguer gravity data of the Gediz Graben and surrounding area in Western Anatolia and its tectonic implications, *Journal of The Balkan Geophysical Society*, Vol. **15**, No. 2, p. 19-30.

[4] D . B haskara Rao, 1986, Modelling of sedimentary basins from gravity anomalies with variable density contrast, *Geophys. J. R. astr. Soc.* **84**, 207-212

[5] Emilson Pereira Leite and Naomi Ussami, 2006, Linear inversion of a negative gravity anomaly in

Se Rio Grande cone:a graben on oceanic crust, *Revista Brasileira de Geofˊısica*, **24**(3): 429-439

[6] Kamtono and D. D. Wardhana, 2012, Nose Structure Delineation of Bouguer Anomaly as the Interpretation Basis of Probable Hydrocarbon Traps: A Case Study on the Mainland Area of Northwest Java Basin, *Indonesian Journal of Geology*, Vol. **7** No. 3: 157-166

[7] Blakely, R. J., 1996, *Potential Theory in Gravity and Magnetic Applications*. Cambridge: Cambridge University Press.

Influence of wind and topography on distribution of H_2S concentration at Cirata hydropower plant and its impact on corrosion rate

Sri Suryaningsih, Jajat Y. Mindara
Department of Physics, Padjadjaran University, Jl. Raya Bandung-Sumedang Km. 21, Jatinangor 45363, Sumedang, West Java, Indonesia

E-mail: sri@phys.unpad.ac.id

Abstract. Emission of hydrogen sulfide (H_2S) gas by diffusion and by bubbling into the atmosphere at tailrace of Cirata hydropower plant give an environmental impact on surrounding area, especially on corrosion of metal surfaces. Field study shows that the highest H_2S concentration and the highest corrosion rate are located at the location of Tailrace, with an average concentration of $152.63\mathbf{x}10^{-3}$ $\mu g.m^{-3}$, and corrosion rate of 0.5010 $mm.yr^{-1}$. Computational Fluid Dynamics is used to develop a simulation model that replicates the function of wind in the distribution of H_2S concentration over the complex three dimensional terrain. Results from this simulation are presented and compared with field study data to determine its impact on corrosion rate.

1. Introduction

Sediment transport along the Citarum river permanently accumulates at the end of the reservoir bed of the Cirata dam. Fishery productivity in the reservoir which produces sediment and nutrient does also directly and indirectly contribute to the accumulation of sediment. This sediment contains dissolved hydrogen sulfide (H_2S) gas which is produced by the decaying process of organic material [1] from the sediment.

Hydropower activities will create downstream of sediment, leading sediment to the penstock (water tunnel), turbine and tailrace (water outlet). Turbulence that occurs will produce bubbles and droplets which will increase the SSA (specific surface area) significantly [2]. As an air-water interface, SSA will emit H_2S into the air. In aerobic condition H_2S will be oxidized to H_2SO_4 , as an catalytic compound in the process of electrochemical corrosion [3] that will increase corrosion rate significanly.

The main purpose of this work is to analyze how H_2S concentration distributes induced by wind and topography in the study area during 10 August 2013 until 11 November 2013. Computational Fluid Dynamics (CFD) software is used to generate simulated imagery inside the domain of interest, and compare this result with field study data.

2. Materials and Methods

In Figure 1, a rectangular region of the study area is located by four corners (107°19'40.18"E; 6°39'48.96"S; 107°22'12.00"E; 6°42'24.48"S) , and covering about 22.4 square kilometers. H_2S, wind sample and placement of corrosion test coupons were observed and collected from four locations within boundary region: • A (Residencial, 6°40'13.19"S- 107°20'58.81"E); • B (Intake, 6°41'24.96"S-107°20'45.86"E); • C (Tailrace, 6°40'53.94"S-107°20'52.00"E); • D (Office complex, 6°42'12.60"S-107°21'8.47"E).

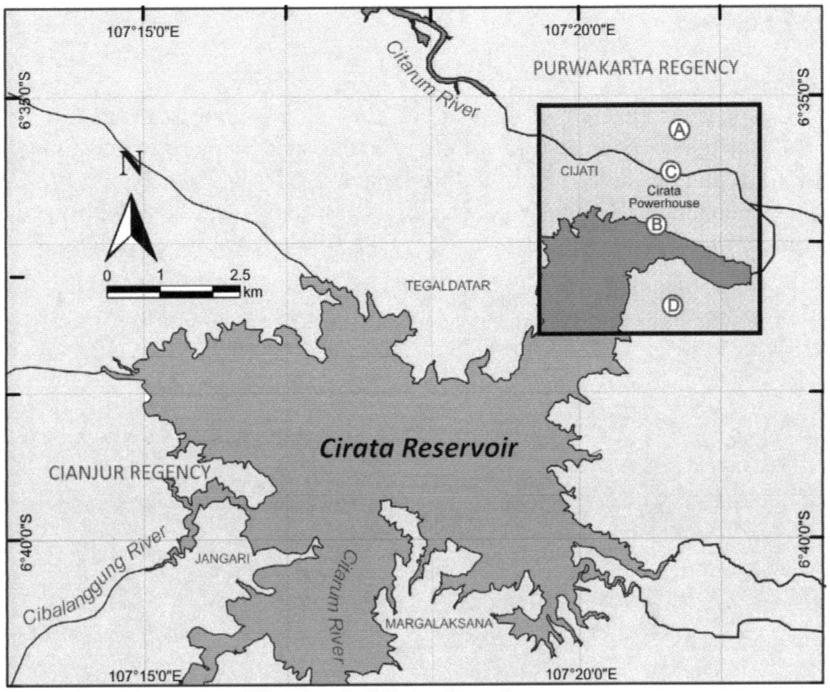

Figure 1. Map of study area.

Three dimensional topography was used in this work to simulate distribution of H_2S concentration using CFD. Digital Elevation Models (DEM) data were extracted form Google Earth at boundary region. The procedure of extracting includes: determining the region of the study area first, specifying the number of points to be extracted by a grid with given rows and columns, extracting the elevation of points column by column (x), tracing the elevation profiles (y,z), finally export point cloud data (x,y,z) into CFD to create 3-D geometry model.

Hourly wind data were extracted from online weather forecast station provided by Norwegian Meteorological Institute (http://www.yr.no), which is the data supplied by the European Organization for the Exploitation of Meteorological Satellites (http://www.eumetsat.int/website/home/index.html). This station was chosen because data from this station closely similar to the sample of wind data from in situ measurement, beside avoiding extrapolation caused by local effects such as the height of an anemometer, relief of the terrain, the land cover and other objects.

Air samples were observed and collected for hydrogen sulfide at four location (A, B, C and D) within the boundary region. The test was based on Standar Nasional Indonesia SNI 19-4844-1998 [4].

Corrosion test was using Carbon Steel Plate Coupon type SS400 from Krakatau Steel that has the following chemical composition by mass%; C (0.007%), Mn (0.311%), S (0.013%), P (0.006%), Si (0.018%), Cr (0.0097%), Mn (0.311%), A (0.039%), Cu (0.039%), Mo (0.021%), Ni (0.023%, V (0.00065%) and the rest is Fe (99.5%). Test coupons are individually mounted on a rack that were electrically insulated from surrounding and making an angle of 45° towards the horizontal plane.

These 24 pieces of test coupons are on average 40 mm wide, 60 mm long, and 1 mm thick and exposed outdoor during study period at four location (A, B, C and D) within the boundary region. After completion of exposed period (2250 hours), corrosion rate were calculate using weight loss method. The test was based on American Society for Testing and Materials (ASTM) G50-76, Standard Practice for Conducting Atmospheric Corrosion Tests on Metals (ASTM G50) [5].

The Comsol Multiphysics was used in this study. The computational model solves the steady-state Navier-Stokes equations which are formulations of mass, momentum and energy conservation laws for fluid flows. The fluid flow is described by the Navier-Stokes equations and the mass flux is given by diffusion and convection.

3. Results and Analysis

Wind speed and direction

The total amount of wind events collected during the study period is about 750 events (calm wind event: 18.86%). The northerly winds blow during the day and southerly winds blow during the night. Figure 2 shows that the northerly winds are predominantly from the N-NNE quadrant and southerly winds from S-SSE quadrant. Highest wind speeds are normally associated with winds from the NNE. In general, average wind speed is about 2 $m.s^{-1}$ for both south and north direction.

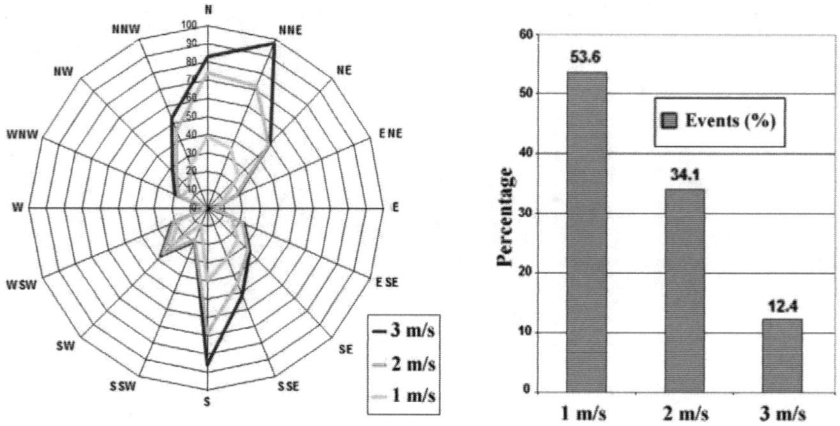

Figure 2. Wind direction (left-hand side) and distribution of wind speed (right-hand side).

H_2S concentration and corrosion rate

H_2S is heavier than air, it will tend to accumulate near the ground when leaking into the atmosphere. H_2S gas is oxidized to H_2SO_4 by thiobacillus bacteria living on the surface near the ground [6]. This acid subsequently condenses along the surface and creates significant corrosion to the iron surface exposed to the air.

Table 1. Field measurement of H_2S concentration and corrosion rate.

Location	Hydrogen Sulfide (H_2S) Measurement						Corrosion Rate Measurement				
	I	II	III	IV	Average		I	II	III	IV	Average
	$\mu g.m^{-3}$	$\mu g.m^{-3}$	$\mu g.m^{-3}$	$\mu g.m^{-3}$	$\mu g.m^{-3}$	$mol.m^{-3}$	$mm.yr^{-1}$	$mm.yr^{-1}$	$mm.yr^{-1}$	$mm.yr^{-1}$	$mm.yr^{-1}$
A	9.46	2.67	6.06	0.40	4.65	$3.39 \cdot 10^{-7}$	0.0181	0.0323	0.0379	0.0542	0.0356
B	9.46	2.67	1.53	2.67	4.08	$2.98 \cdot 10^{-7}$	0.0335	0.0335	0.0350	0.0480	0.0375
C	8.33	516.50	81.89	3.79	152.63	$1.11 \cdot 10^{-5}$	0.2203	0.4905	0.6419	0.6512	0.5010
D	4.93	3.80	n/d	3.79	4.17	$3.04 \cdot 10^{-7}$	0.0375	0.0637	0.0626	0.0536	0.0543

The concentration of H_2S and corrosion rate in location C is very high, far exceeding the three other locations (Table 1). This is because location C is a pollutant source, and the three other locations (A, B and D) can be assumed as dissemination areas.

Simulation of wind and H_2S concentration

North wind and south wind were chosen because they represent the typical wind direction in this area. Wind speed of 2 $m.s^{-1}$ was chosen to present the average ability of wind speed to move the pollutant across this area. A cross section at location C (at 107°20'52.00"E, from 6°39'48.96"S to 6°42'24.48"S) was chosen because it represents the pollutant source, and also parallel and closely colinier to other cross section at location A, B and D.

Because H_2S is heavier than air, the wind near the surface is very influential in the spread of H_2S concentrations. Figure 3 shows the wind contour near the surface in a simulation with a north wind. Location C and B are lower in altitude and both getting wind with low speed (0 to 1 $m.s^{-1}$) which is potential for pollutants to be trapped inside. This condition is not much different with a south wind simulation. Noted that the models used in the simulation below (Figure 3, Figure 4, Figure 5) are scaled down by 10 for saving computation time.

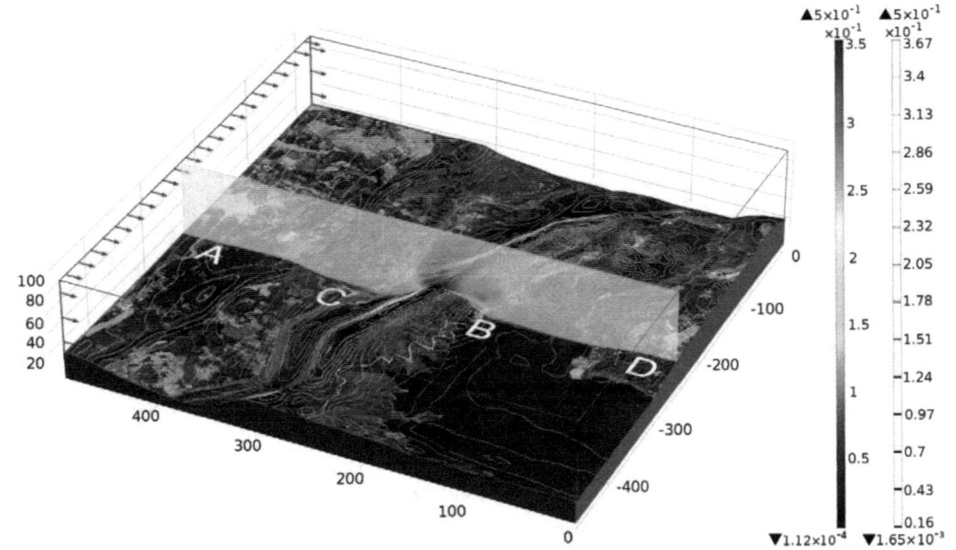

Figure 3. Contour of near surface wind.

North wind simulation.

In the figure 4, it shows two basins that could cause pollutants to be trapped, at the location C where the H_2S gas emissions happen and location B above the surface reservoir. Wind speed in both locations plummet

Most of the H2S gas was trapped in the south walls of the basin (location C). Some of 8.65x10-7 mol.m-3 H2S concentrations trapped above the surface reservoir (location B) due to turbulence at low wind speeds. An amount of 1.96x10-7 mol.m-3 H2S concentration will reach the location D, by distance of more than 2,500 meters from the source.ed range between 0-1 $m.s^{-1}$.

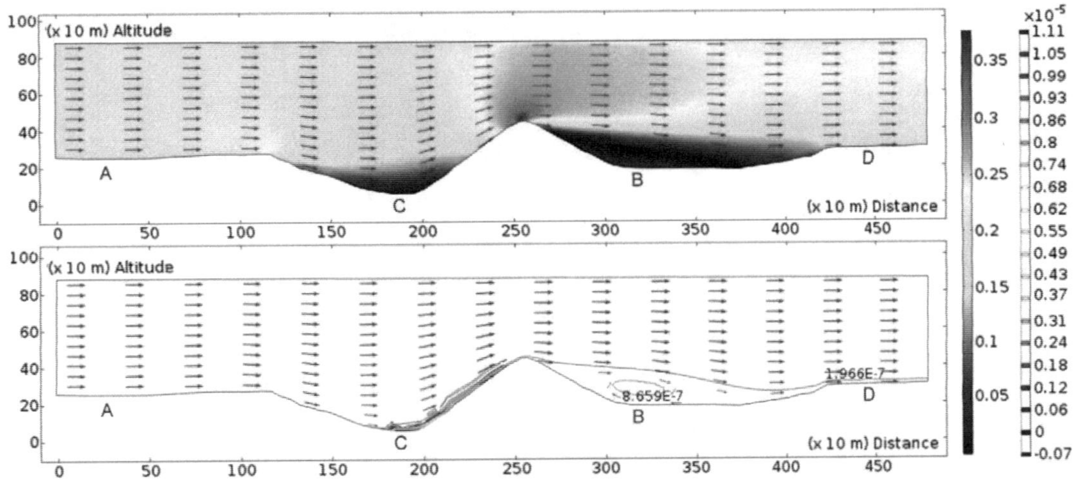

Figure 4. Wind magnitude, wind direction and distribution of H_2S concentration north wind simulation.

South wind simulation.
Figure 5 shows that wind blows with low speed (0 to 1 m.s^{-1}) at the location C and location A.

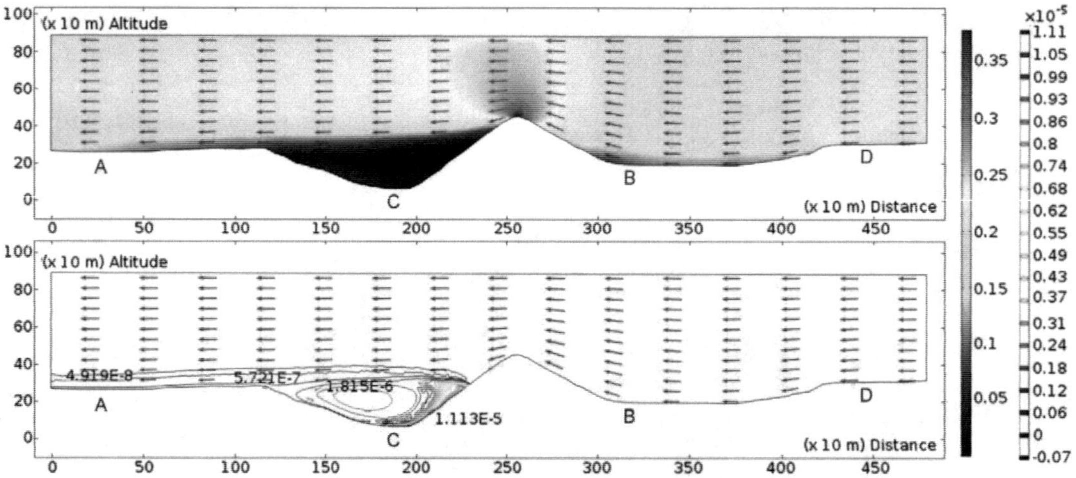

Figure 5. Wind magnitude, wind direction and distribution of H_2S concentration south wind simulation.

In detail, turbulence with small wind speeds occurs in the basin (location C). Most of the H_2S gas will be trapped in the south walls of the basin. An amount of 4.91x10^{-7} mol.m^{-3} H_2S concentrations will reach the location A, by the distance of more than 1,500 meters from the source.

H_2S concentration and corrosion rate
From the results presented in Figure 6 it is apparent that the simulation of H_2S concentration will lead to a small deviation compared with H_2S concentration from measurement result. This is becouse the wind conditions are changing at the time of sampling H_2S, but in general the results showed the same pattern.

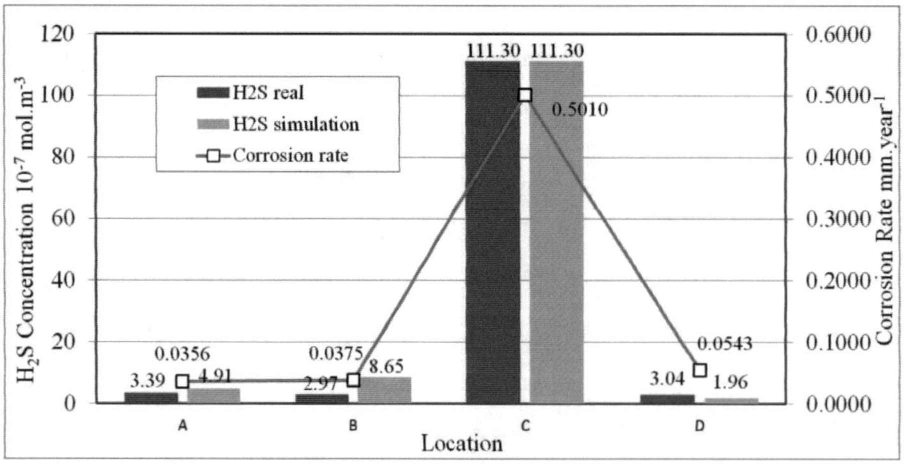

Figure 6. Measured and simulated H_2S concentration and corrosion rate.

Corrosion rate also showed a similar pattern as the H_2S concentrations, but there was some difference in location D. In this case, the corrosion rate is slightly higher than at location A and B, whereas the level of H_2S concentration in the location D is the lowest. This is probably caused by other physical parameters that affect the corrosion rate (such as dew point temperature, moisture, corrosive gas, and other particulate matter) that are not included in this study.

4. Conclusion

Wind direction is predominantly from the N-NNE quadrant during the day, and from S-SSE quadrant during the night. Highest wind speeds are normally associated with winds from the NNE. In general, average wind speed is about 2 $m.s^{-1}$ for both south and north direction.

Low level of wind speed, wind direction and topography with a high slope angle lead to high concentration of H_2S gas trapped in the basin, only less than 8% of H_2S concentration spread across the hill. Consequently, high concentration of H_2S gas in the basin will increase corrosion rate in this area significantly.

Field study of the H_2S concentration and corrosion rate illustrates the same pattern as the spread of H_2S concentration using a CFD simulation. All the parameters are at the highest level at the tailrace and three other locations are lowest, hence the presumption that the spread of H_2S concentration derived from the location of Tailrace as a source, and three other locations as dissemination areas could be proven.

References

[1] Shreir L L 2000 *Corrosion: Metal / Environment Reactions* 3rd Ed. (Oxford:Butterworth Heinemann.) p 2:35

[2] Chanson H 1997 Measuring Air-Water Interface Area in Supercritical Open Channel Flow, *Water Research.,* IAWPRC, Vol. **31**, No. 6, p 1414-1420

[3] Ahmad Zaki 2006 *Principles of Corrosion Engineering and Corrosion Control* (Oxford: Butterwoorth-Heinemann) p 125, 374, 555, 558

[4] Standar Nasional Indonesia 1998 *Test Methods for hydrogen sulphide (H_2S) concentration in air using spectrophotometer air samplers (HVS)* SNI 19-4844-1998 Badan Standarisasi Indonesia

[5] American Society for Testing and Materials 1992 G50-76 *Standard Practice for Conducting Atmospheric Corrosion Tests on Metals* Annual Book of ASTM Standards (Philadelphia, ASTM) p 371-374, 201-205

[6] Cullimore D Roy 2010 *Practical Atlas for Bacterial Identification* 2nd Ed. (Boca Raton : CRC Press) p 46, 94, 133, 139

IOP Conf. Series: Earth and Environmental Science **29** (2016) 012015 doi:10.1088/1755-1315/29/1/012015

Magnetic Properties of Soils from Sarimukti Landfill as Proxy Indicators of Pollution (Case Study: Desa Sarimukti, Kabupaten Bandung Barat)

Gesti Cita Novala, Dini Fitriani, Kusnahadi Susanto, Kartika Hajar Kirana
Study Program of Geophysics, Universitas Padjadjaran,
Jalan Raya Bandung-Sumedang km.21, Sumedang 45363, Indonesia

Email: dini@geophys.unpad.ac.id

Abstract. Leachate is the liquid arises from waste disposal. It contains heavy metals and magnetic minerals. Leachate could penetrate into sub surface that cause soil contamination. We have studied magnetic properties of soils from three zones in the Sarimukti landfill. We measured magnetic susceptibility in dual frequency and temperature dependent of susceptibility. The results showed that magnetic susceptibility at low frequency (χ_{LF}) of soil samples have the value ranging from 50-1400 (x 10^{-8} m^3/kg). It infers that the samples were dominated by ferrimagnetic minerals. The relative difference of magnetic susceptibility measured at two frequency ($\chi_{FD}(\%)$) are less than 4%. It indicates that magnetic minerals probably derived from anthropogenic sources. There is negative correlation between χ_{LF} and $\chi_{FD}(\%)$. It supports the analysis that the source of magnetic minerals is anthropogenic. The results of the temperature dependent magnetic susceptibility measurements showed that the peak of magnetic susceptibility in heating curve occurs at the temperature about 200°C and 500°C. After 500°C, heating curve rapidly decreases. It can be inferred that the dominant type of magnetic minerals in the samples is titanomagnetite.

1. Introduction

Leachate is a liquid derived from waste which has yellow, brown or black colour [1]. Leachate is produced when water in the form of rainfall or groundwater percolate into the waste [2]. Leachate will be found around landfills. The large volume of incoming waste and heavy rainfall in the landfill area is potentially generating leachate in large numbers.

Leachate can penetrate into subsurface and this can contaminate soil and groundwater. Leachate of the landfill contains heavy metals, magnetic minerals, hazardous chemicals, bacteria, and viruses. Therefore, it is necessary to identify the contamination of the soil to provide a solution to the problem of soil contamination caused by leachate.

In this research, we identified of magnetic properties on soil samples taken from the Sarimukti landfill, Kabupaten Bandung Barat. Magnetic measurements performed in this study are the measurement of the frequency dependent magnetic susceptibility and temperature dependent magnetic susceptibility. The aim of the study is to identify the magnetic properties of the soil in the Sarimukti landfill and surrounding area as an indicator of soil contamination.

2. Methods

We have collected soil samples from three zones of Sarimukti landfill, namely zone 1, zone 2, and zone 3. Zone 1 and zone 2 are inactive zones, while zone 3 is an active zone. Inactive zone is a zone that is no longer used as landfill, otherwise active zone is a zone that is still active as a landfill and produce landfill leachate. The three zones have different topography as shown in Figure 1.

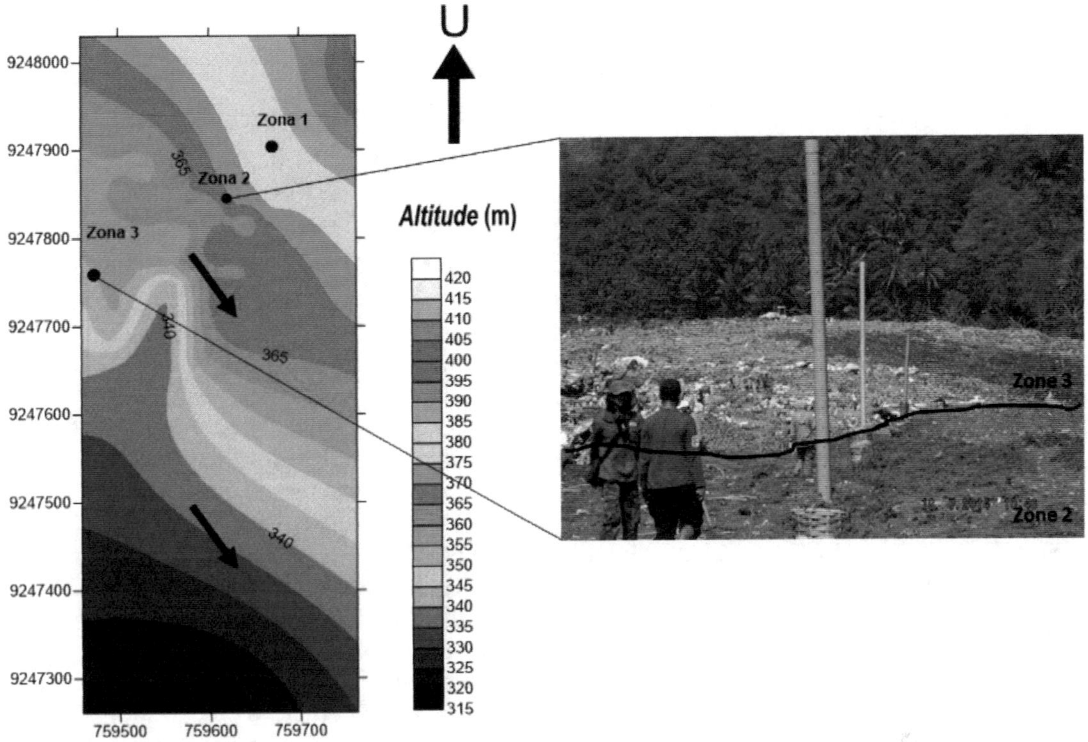

Figure 1. Topographic map of Sarimukti landfill. The arrows show the ground water flows around the target area [3] and the picture show the boundary between the zones.

Soil samples were obtained from surface to a certain depth. The samples were then placed into 10 cm^3 sample holder. In total, we have 25 samples. All samples were subjected to magnetic susceptibility measurement by using Bartington Magnetic Susceptibility system with sensor B (Bartington MS2B). MS2B operates at low frequency (470 Hz) and high frequency (4700 Hz). The relative difference of magnetic susceptibility values in these frequencies is known as frequency-dependent susceptibility, χ_{FD} (%). We also measured temperature-dependent susceptibility for some selected samples using Bartington MS2 with sensor W.

3. Results and Discussion

In zone 1, we have samples from 0-60 cm depth. Magnetic susceptibility value at low frequency (χ_{LF}) ranges from 1150-1400 (× 10^{-8} m^3/kg). The highest value of χ_{LF} is at depth of 30-40 cm. Soil samples from zone 2, taken from 0-50 cm depth, has χ_{LF} ranging from 800-1000 (× 10^{-8} m^3/kg) with the highest value is at depth of 20-30 cm. In zone 1 and zone 2, there is similiar pattern of variation in magnetic susceptibility with depth.

Soil samples from zone 3 has lower values of χ_{LF} than those of zone 1 and zone 2. In zone 3, the χ_{LF} varies from 50-550 (× 10^{-8} m^3/kg). The highest value of χ_{LF} is at topsoil. Plot of variation of magnetic susceptibility values with depth is shown in Figure 2.

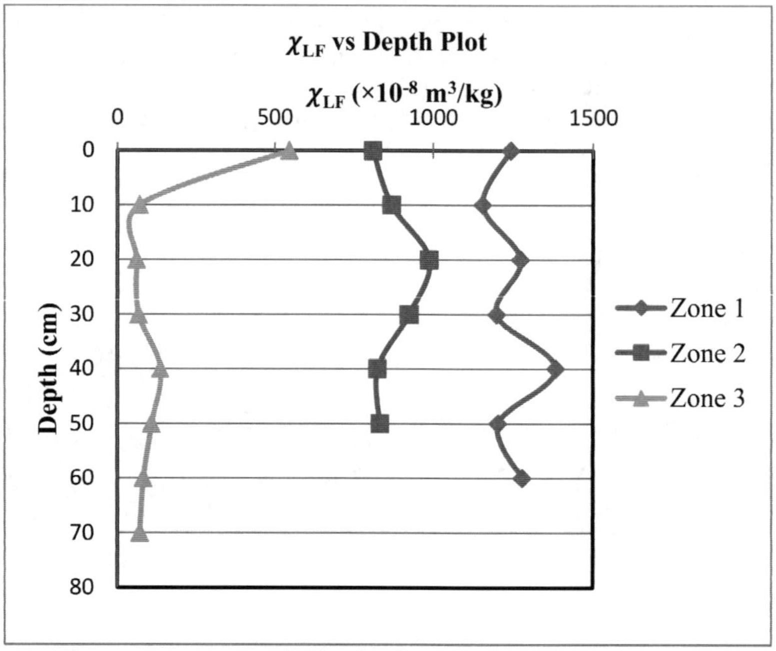

Figure 2. χ_{LF} vs Depth Plot for Samples in Sarimukti Landfill.

Samples of rocks and soils, that dominated by silty clay [3] showing purely paramagnetic behaviour rarely show χ_{LF} values exceeding 10×10^{-8} m^3/kg. Therefore, as a rule-of-thumb, the χ_{LF} of any sample with a value less than this is probably controlled by the concentration of paramagnetic minerals and for values greater than this by ferrimagnetic minerals [4]. Based on χ_{LF} value; it can be deduced that magnetic minerals contained in the samples are dominated by ferrimagnetic minerals.

The presense of magnetic minerals can occur due to soil formation process (pedogenic) or also due to other sources, such as human activities (anthropogenic). Anthropogenic sources can be industrial activities, household or vehicle derived particulates.

The pattern that high magnetic suscetibility values decrease from the surface to a certain depth indicates a large contribution of anthropogenic related ferrimagnetic minerals [5]. This pattern occur in zone 3. It indicated that magnetic minerals in this zone derived from anthropogenic sources.

The χ_{FD} (%) values of the samples in landfill area are less than 4%. Low values of χ_{FD} (%) (1-4%) are commonly found in contaminated soils. Magnetic minerals in soils derived from pedogenic process has a higher value (~10%) [6]. It infers that the magnetic minerals from study area are derived from anthropogenic sources.

Statistical analysis of χ_{LF} and χ_{FD} (%) shows negative correlations. The existence of negative correlations. The exixstence of negative correlation between χ_{LF} and χ_{FD} (%) indicate that the magnetic minerals are from anthropogenic sources. It supports the analysis that the source of magnetic minerals is anthropogenic. Plot χ_{FD} (%) vs χ_{LF} is presented Figure 3.

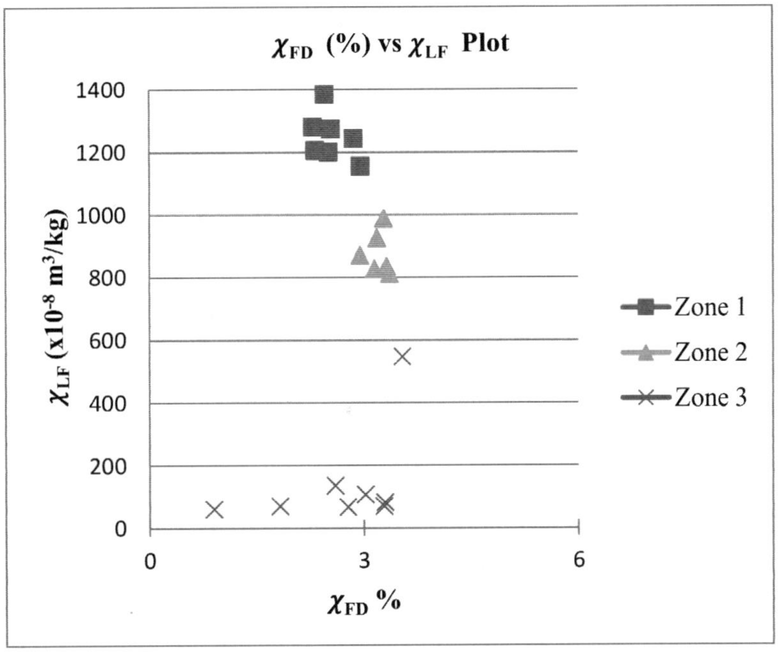

Figure 3. χ_{FD} (%) vs χ_{LF} Plot for Samples in Sarimukti Landfill.

Temperature-dependent susceptibility analysis is widely used for determining the Curie point [7] at which ferrimagnetic or ferromagnetic mineral becomes paramagnetic during heating. Temperature-dependent susceptibility measurements were performed by heating the samples to 700°C and cooling it to room temperature. Cooling process is often used as an indicator to identify mineralogical phase changes. In this study, we selected one sample in each zone to be subjected to temperature-dependent susceptibility measurement.

Sample from zone 1 showed an increase in magnetic susceptibility as temperature increase with the peak at about 200-300 °C. Heating curve rapidly decrease after 500°C, as shown in figure 4.

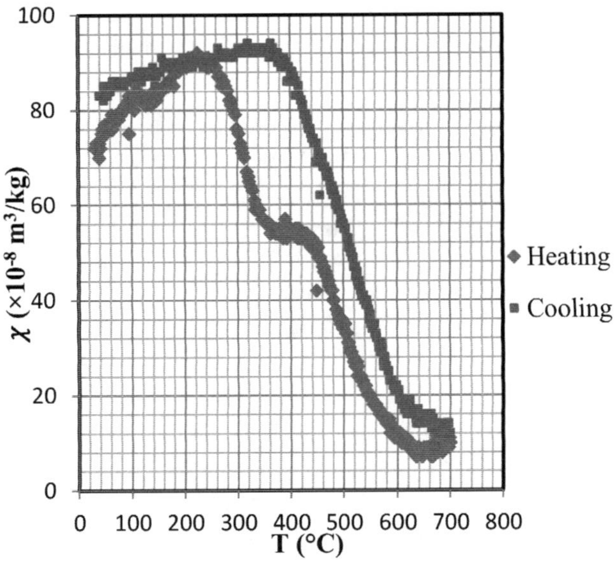

Figure 4. Temperature vs Magnetic Susceptibility Plot for Zone 1.

Peak increases in magnetic susceptibility from zone 2 at about 200-300 °C. After the 500°C heating curve is rapidly declining. Plot of temperature on the magnetic susceptibility for zones 2 are presented in Figure 5.

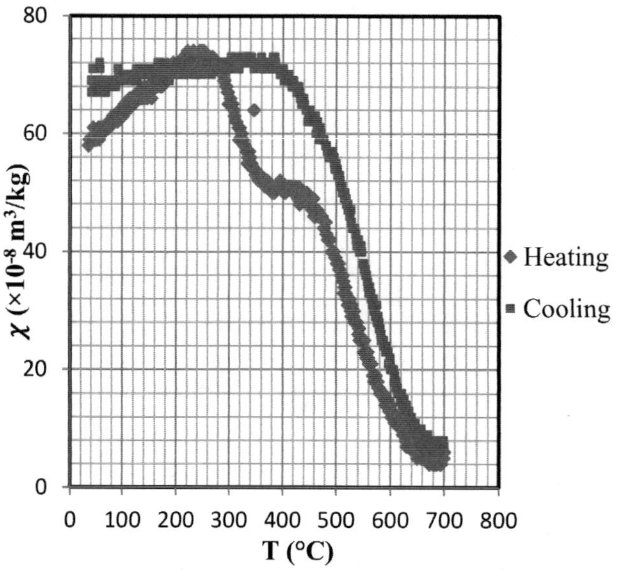

Figure 5 . Temperature vs Magnetic Susceptibility Plot for Zone 2

Based on plot of temperature with magnetic susceptibility values from zone 3, magnetic susceptibility increases at 200-300 °C when the sample was heated. As same as previous samples, after 500°C heating curve is rapidly declining, as shown in figure 6.

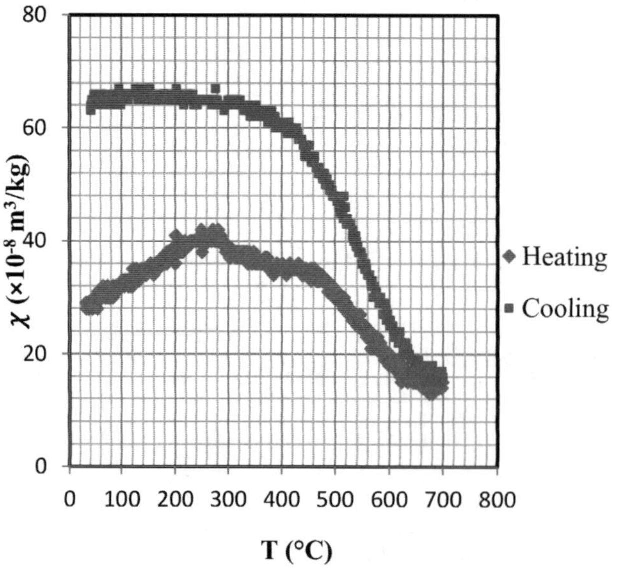

Figure 6. Temperature vs Magnetic Susceptibility Plot for Zone 3.

The presence of magnetite phase is at a temperature of about 580°C. This shows that the Curie point of magnetite to be around 580°C or more precisely ranged from 575-585 °C. All samples has heating that curve decreases after 500°C. It indicates the occurrence of phase transitions. Based on these results, three samples have similar mineralogical phase with the mineral magnetite. In addition, the peak increase in any samples taken from each zone occurs at about 200°C and 500°C. Therefore, the dominant type of magnetic minerals in the sample is titanomagnetite.

4. Conclusions

1. Based on magnetic susceptibility value, magnetic minerals in the samples are dominated by ferrimagnetic.
2. Source of magnetic minerals are anthropogenic.
3. According to temperature dependent susceptibility measurement, the dominant type of magnetic minerals is titanomagnetite.

References

[1] Zouboulis AI Chai X L Katsoyiannis I A 2004 The application of bioflocculant for the removal of humic acids from stabilized landfill leachates. *Journal of Environmental Management* **70** 35–41.

[2] Christensen T H Kjeldsen P Bjerg P L Jensen D L Christensen J B Baun A Albrechtsen H J Heron G 2001 Biogeochemistry of landfill leachate plumes *Applied Geochemistry* **16** 659-718.

[3] Suganda, B.R., T.Y.W.M. Iskandarsyah, M.S.D. Hadian. 2012. Prediksi arah pencemaran air tanah akibat pembuangan sampah akhir di daerah Sarimukti dan sekitarnya Kabupaten Bandung Barat, Provinsi Jawa Barat. *Bulletin of Scientific Contribution,* **10** 31-41.

[4] Dearing J A 1999 Environmental Magnetic Susceptibility Using the Bartington MS2 System *British Library Cataloguing in Publication Data.*

[5] Baghdadi E M Barakat A Sajieddine M Nadem S 2011 Heavy metal pollution and soil magnetic susceptibility in urban soil of Beni Mellal City (Morocco) *Environ Earth Sci* DOI 10.1007/s12665-011-1215-5.

[6] Bijaksana S and Huliselan E K 2010 Magnetic properties and heavy metal content of sanitary leachate sludge in two landfill sites near Bandung, Indonesia *Environ Earth Sci* **60**:409–419 DOI 10.1007/s12665-009-0184-4.

[7] Lu and Bai 2008 Magnetic characterization and magnetic mineralogy of the hangzhou urban soils and its environmental implications *Chinese Journal of Geophysics* **Vol.51**, No.3, 2008, pp: 549_557.

IOP Conf. Series: Earth and Environmental Science **29** (2016) 012016 doi:10.1088/1755-1315/29/1/012016

VLF -MT Survey around Nakadake crater at Aso Volcano

A Harja[1], W Srigutomo[2], Y Tanaka[3], T Kagiyama[3] and D Sutarno[2]
[1]Geophysics Exploration and Natural Resources Laboratory, Department of Geophysics, Universitas Padjadjaran, Indonesia.
[2]Electromagnetic Induction Lab., Physics of Complex System Div., Fac. Of Mathematics and NaturalSciences, Institut Teknologi Bandung, Indonesia.
[3]Aso Volcanological Lab., Institute for Geothermal Science, Grad.School of Science, Kyoto University. Japan

E-mail: asep.harja@geophys.unpad.ac.id

Abstract. Shallow resistivity structure around Nakadake craters has been revealed by extensive VLF- MT survey from August to October 2006. This structure comprises a varied resistivity value at the upper and bottom layer of the western flank of craters; a conductive upper layer at the fourth Crater; and a resistive upper layer at southern part of craters. The conductive value of second layer at a line in the western flank and southern part of craters at several ten meter depth, interpreted as water saturated, and high resistivity at second line of western flank of craters as sheet of lava and at fourth crater is interpreted as lava rock . At western flank of craters, we found the resistivity value and the depth of second layer decrease in depth with decreasing of distance to craters line.

1. Introduction

Aso Volcano is located on the Kyushu island, in the south-western , part of Japan. Nakadake, one of the youngest volcanoes in the central cone complex of Aso Volcano, has a currently active crater, named first Crater, and other 6 craters in the NNW-SSE direction. The first crater has been ejecting large quantities of ash, mud and scoriaceous ejecta in strombolian type eruptions every years [1], and is also the local of episodes of continuous volcanic tremors which accompany the volcanic activity. The 1st crater diameter is about 400 m. During quiet volcanic states, a crater lake is usually present in the bottom of the crater, the surface temperature of the lake being about 600^0 C. Figure 1 shows the location of Nakadake craters and mapping points of the surveys.

Many active volcanoes bear a hydrothermal fluid system within their edifice. Such fluid easily infiltrates within permeable ground and efficiently transports heat because of its large heat capacity whichever the origin is precipitation from the atmosphere or magmatic volatile. Electrical resistivity is one of the most useful methods to examine the thermal process taking place [2], because the electrical resistivity of the ground highly depends on the temperature and the concentration of water and volcanic gas in the permeable porous medium that usually composes the volcanic edifice [3].

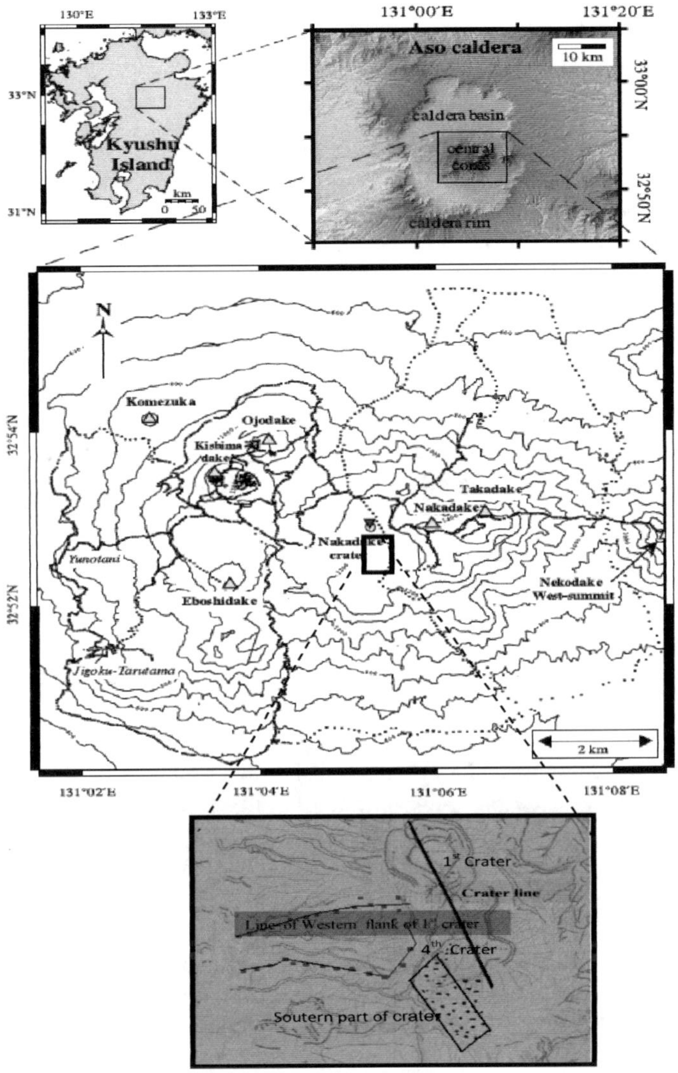

Figure 1. Location map of the study field. red and blue dots show the measurement points of VLF-MT.

In this paper, we carried out VLF-MT survey to infer the shallow resistivity structure around the craters and discuss the hydrothermal system beneath crater rim of Nakadake craters. Interaction between subsurface water and heat will be discussed in relation to conductive structure. VLF signals (15-30 kHz) are used to study the variation of the electrical conductivity in the upper few hundred metres of the Earth's crust. The variation in EM field components relate to the electrical structure of the subsurface. The VLF resistivity mode (VLF-MT), which also utilizes electric field, was proposed by Collet and Becker [4]. The VLF method maps shallow depth inhomogeneity and widely used in mineral exploration, groundwater exploration and waste management and in structural studies.

VLF-MT measurements are used to record the horizontal electric field component and an orthogonal horizontal magnetic field. Because of large distance from the VLF transmitter, the primary EM fields can be regarded as plane waves. It is possible to use either the E- or H- polarization. For E-polarization, where electric field is parallel with the geological strike, apparent resistivity, ϱ_a, and impedance phase, ϕ, computed from the horizontal electric field E_y and magnetic field H_z are given by the formula:

$$\rho_a = \frac{1}{\mu\omega}\left|\frac{E_x}{H_y}\right|^2 = \frac{1}{\mu\omega}|Z|^2 \quad (1)$$

and

$$\emptyset = arctan\left[\frac{im(E_x/H_y)}{Re(E_x/H_y)}\right] \quad (2)$$

where μ=μ0 is the magnetic permeability of free space and, ω is anguler frequency of measurement, and Zxy is the surface impedance. The skin depth, δ, is defined as the depth of penetration at which the electromagnetic field reduces to 1/e of its value at the surface of that medium, it is given by :

$$\delta = \left(\frac{2}{\omega\mu\sigma}\right)^{1/2} (unit\ \ length) \quad (3)$$

where σ is conductivity. In practice, Observed VLF-MT data are represented by apparent resistivity, ϱa, and impedance phase, ϕ. Field component Ex in eqs.(1) and (2) is obtained by applying the finite element technique with the Galerkin process directly to Maxwell's equations; the other field components can be computed from Maxwell equations by numerical differentiation. The very fast simulated annealing (VFSA) approach is used as the procedure of optimization in this inversion [5].

2. Very Low Frequency – Resistivity (VLF-MT) surveys.

The mapping points are located on western flank of crater and southern part of craters. The points of VLF-MT survey is shown in Figure 1, line survey on the western flank of the craters, within the fourth crater, and southern part of the craters. As a source is used VLF-MT receiver is VLF radio wave station (JJI; 22.2 kHz) for submarine communications, which located about 100 km SSW away from the present study area. The resistivity structure around crater rim of Nakadake crater was established by VLF (Very Low Resistivity) R or MT surveys along August – October 2006.

The results are presented in three areas that are western flank of craters, southern part of craters and fourth crater. From previous research [1,6], the results of resistivity sounding are composed of fine volcanic ashes for the surface, the second layer is pilled scoriae and the western and southern part of the lowest layer is formed of the high resistive Kusasenri lava rocks. The very low resistivities at the crater of Nakadake (Sunasenri) suggest these regions are composed of scoriae and volcanics ashes. While the high resistive layer found at Umanose and Western slope out the crater, represent the existence of lava sheets.

The area at western flank of craters, has two line surveys, Figure 1. The results, Figure 2, shows the apparent resistivity and impedance phase as function of distance from craters line. It indicates that the various results for three time survey, apparent resistivity tend to decrease and change in impedance phase at same distance from crater line. The second area, Figure 3., is southern part of crater, we have made in contour of apparent resistivity and impedance phase for all data of surveys. These contour show that various apparent resistivity and impedance phase at several locations indicate different resistivity structure of subsurface. The last area is fourth crater, Figure 4, low apparent resistivity and contras in impedance phase contour indicate location of the lower resistivity along the middle of fourth crater.

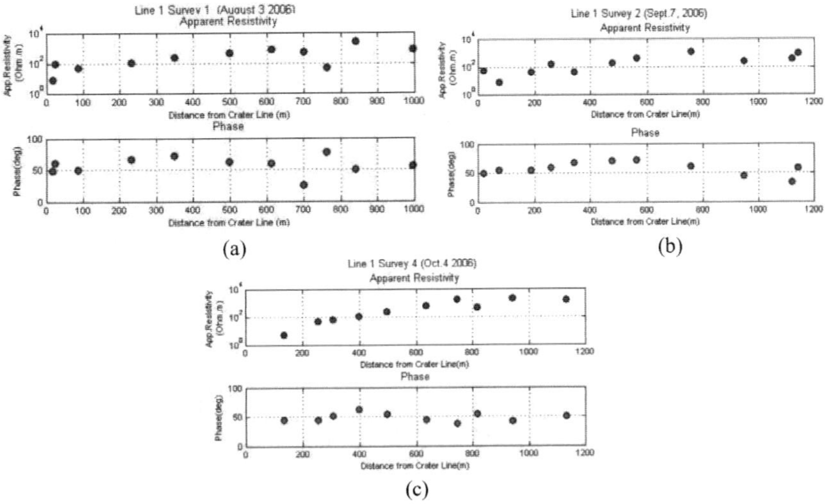

Figure 2. Apparent resistivity and impedance phase of Line 1 of VLF-MT Survey (a) August 3, 2006 (b) Sept. 7, 2006 (c) Oct. 4 2006

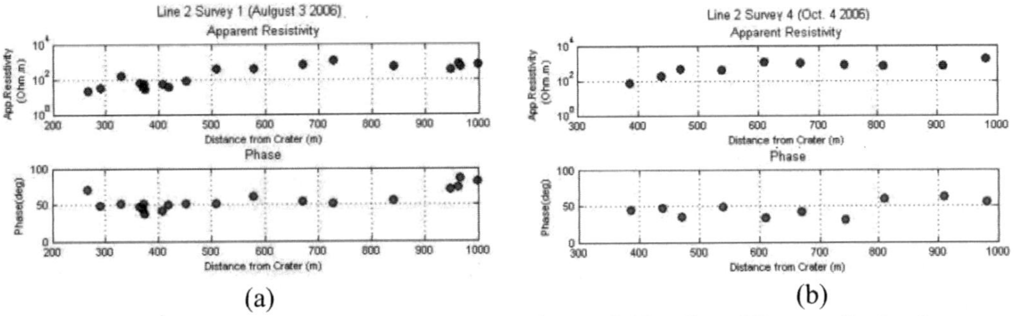

Figure 3. Apparent resistivity and impedance phase of Line 2 at Western flank of craters. Survey (a) August 3, 2006 (b) Oct. 4 2006

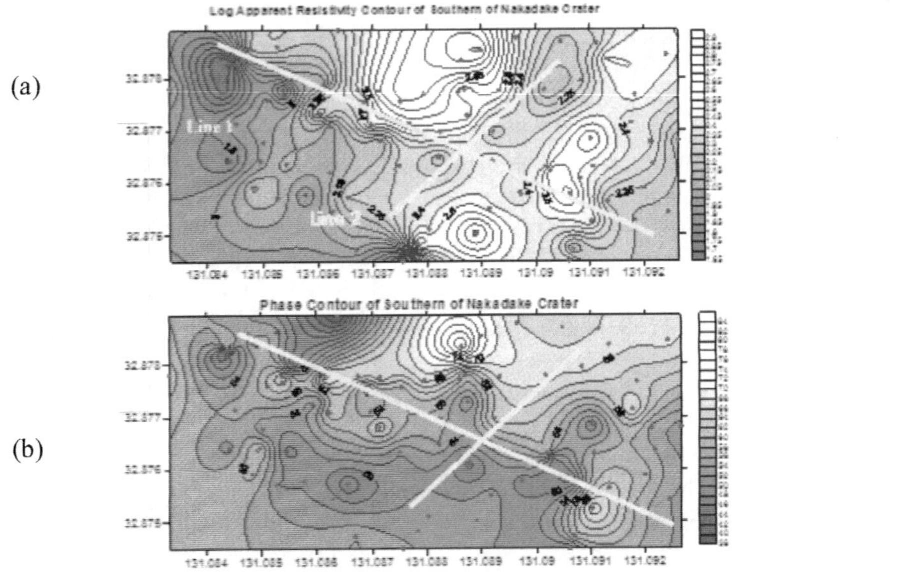

Figure 4. (a). Apparent Resistivity and (b) Impedance phase Contour, at Southern part of Craters

4. VLF-MT Data Inversion and Discussion

The laterally constrained two layer inversion of VLF-MT data algorithm, developed by Pirttijärvi,M [7], was applied to interpret electromagnetic VLF-MT data (apparent resistivity and impedance phase) measured along a single profile at a single frequency. The inversion is made separately for each data point using a one-dimensional two-layer earth model. As such the program suits well the interpretation of the thickness and resistivity of relatively thin resistive (soil and till) layers and the resistivity variations of the basement rocks. Note that the inversion model is still only one-dimensional and that the interpretation results actually yield a pseudosection of the 2-D resistivity distribution.

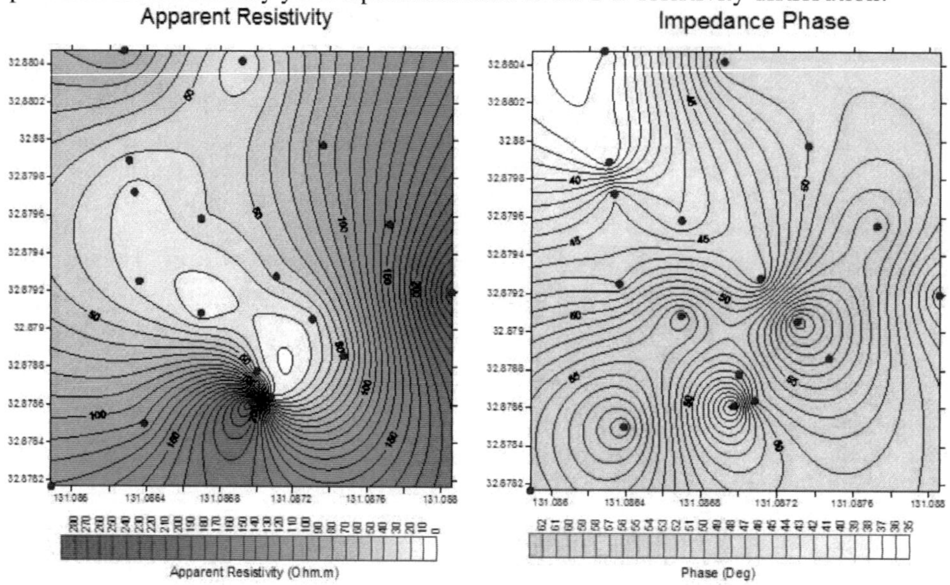

Figure 5. Apparent Resistivity and Impedance phase Contour at fourth Crater

Starting from an initial model linearized inversion with adaptive damping is used to optimize the thickness and the resistivity of the overburden layer and the resistivity of the basement so that the model minimizes the error between the measured and the computed VLF-MT data. Laterally constrained inversion is obtained by minimizing the roughness of the model that is the variation of the model parameters between neighboring points, together with the data error. As a result a smoothly varying (Occam inversion) model is obtained. The roughness and/or the fix/free status the parameters can be set manually to allow discontinuities and to incorporate a priori data in the model.

Source for the VLF-MT measurement use a signal frequency of 22.2 kHz. The quantitative interpretation, equivalent with other geophysical methods, relies on model calculation into the relationship between observed data and real geological condition to yield a quantitative interpretation. The inversion modeling calculates the two-layer model data, using of the measured apparent resistivity. This inversion result shows that the second layer has lower resistivity. It is less than 50 Ωm at 3-10 meters deep and 0-50 meters distant from the crater line (Figure.6a). At the depth of 2-70 meters and the distance of 50 to 1000 meters from the crater, the resistivity is around 50 to 100 Ωm. The inversion result of second survey was convergence with 13,52% for data RMS error and 0.43% model RMS (Figure. 6b) and third survey was convergence with 16.51% for data RMS error and 0.56% model RMS (Figure.6c). Generally, these results show the increasing resistivity of second layer comparing to first survey data.

Generally, the inversion result of third survey shows increasing resistivity value for all , for 125 to 400 meter from crater line resistivity of second layer is 48 to 100 Ωm with 2 to10 m in depth, for the resistivity from 400 to 634 meter is 100-300 Ωm with 10-20 in depth, the resistivity is higher than 300 Ωm for distance from crater more than 634 meter with variation in depth (9 to 50 m). For the second line, the inversion results of first survey was convergence with 12.47% for data RMS error and 0.65% for model RMS. These results show that upper layer is more resistive than the second layer with variation in depth from 13 to 90 meter (Figure 7) , inversely at near crater upper layer more conductive.

For two parallel lines (line-1 and line-2) at 4th crater (Figure 8), the result shows the depth variation of the bedrock (15 to 36 meters). Line-1 of this crater is deeper than eastern part (line-2). From inversion result, the resistivity profile of the crater shows the different structure among line-1 and line-2. The higher resistivity of line-1, compares to line-2, can be interpreted as the domination of lava-rock due to the volcanic activity. Whilst, the lower resistivity is due to the heat below the crater. At the southern part, two perpendicular lines show consistence of resistivity structure, which is more resistive at the upper layer. The decreasing value of the resistivity near the crater line is influence of the volcanic underneath the crater.

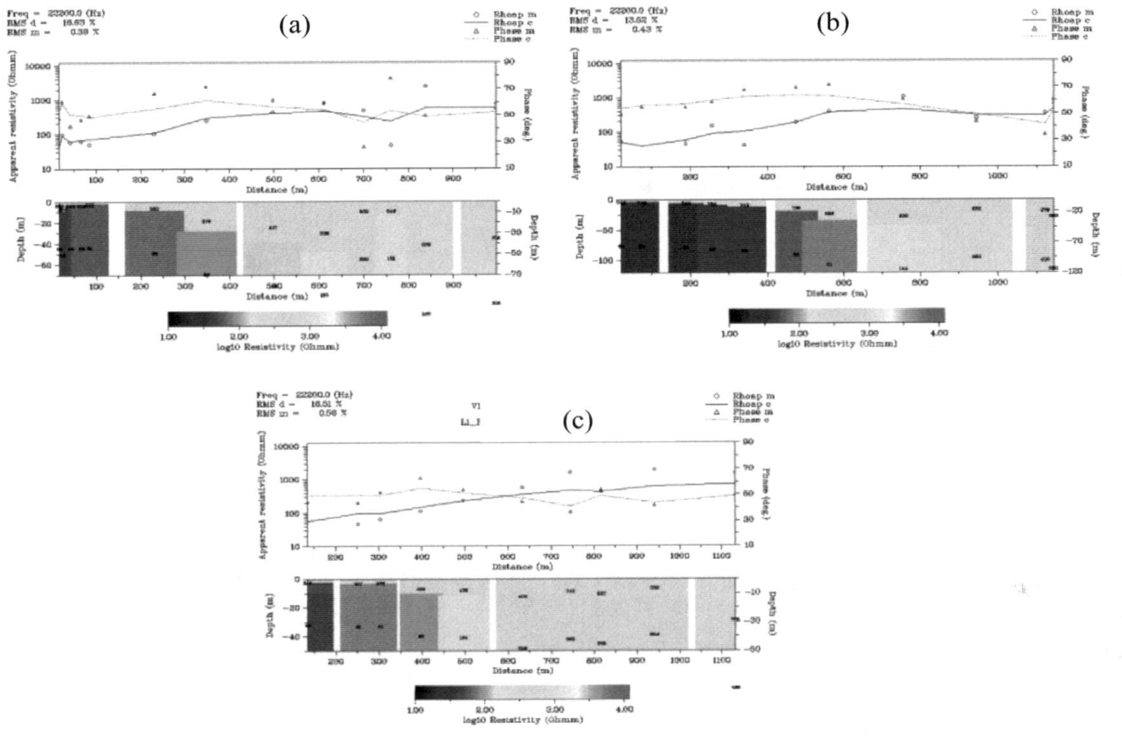

Figure 6. Inversion results of first line at western flank of craters (a) Survey at August 6, 2006; (b) Survey at September 7, 2006; (c) Survey at October 4, 2006

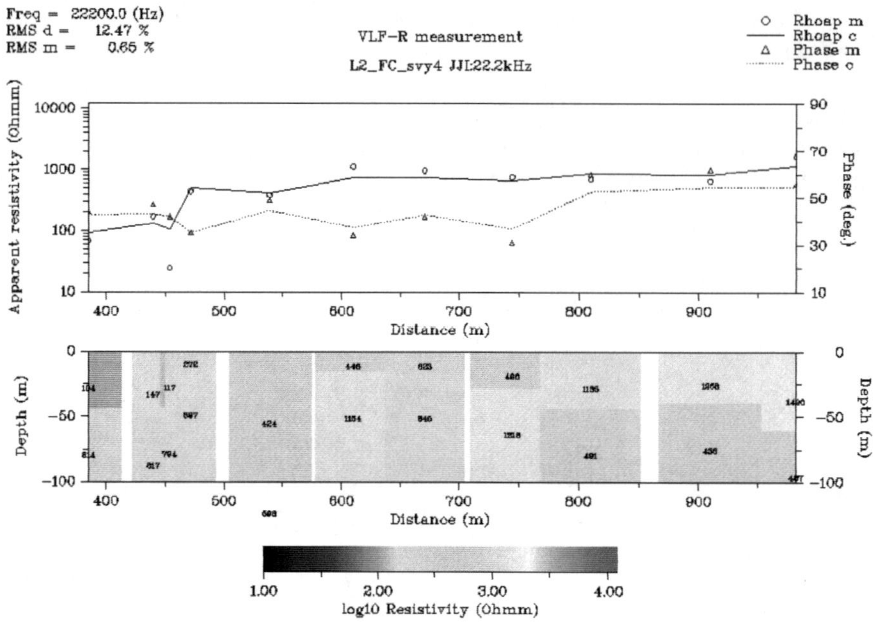

Figure 7. Inversion results of second line at western flank of craters

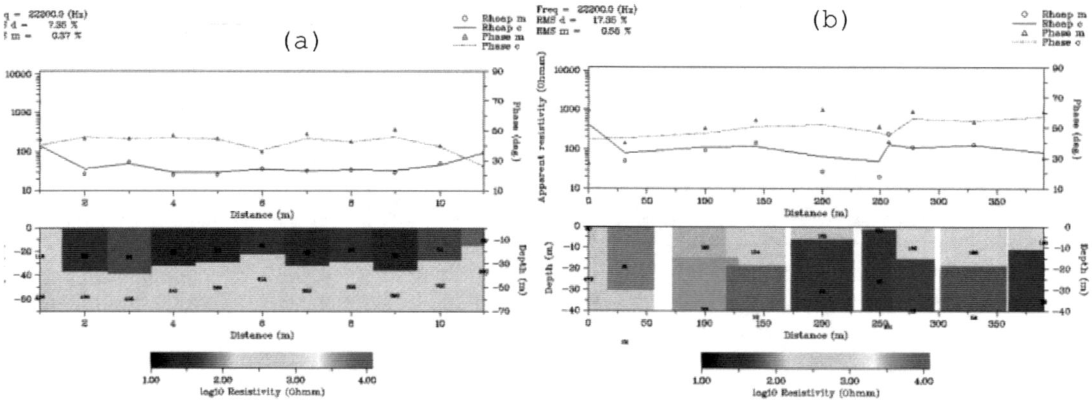

Figure 8. Inversion results of fourth crater (a) line 1 (b) line 2

5. Conclusions

In this study, inverted resistivity revealed general futures of upper and bottom layer. It can be conclude that at western flank of craters has higher resistivity second layer indicate lava sheet and the increasing of resistivity for second survey is caused decreasing groundwater as dominant factor of resistivty value of rock, inversely, for line 1 more conductive of second layer lead to suggestion that shallow subsurface of few meters in depth indicated is repeated heated and supplied by groundwater. Phenomena low resistiviy of surface at fourth crater reveal dominated lava rock of second layer. For upper layer of fourth crater, low resistivity in the top layer thicker is suggested by heat from beneath the crater. The phenomena low resistivity of upper layer in southern part of crater is suggest by water saturated layer through interaction between magma, juvenile gas, and groundwater.

Acknowledgments

The authors wish to express sincere thanks to Aso Volcanological Laboratory, Institute for Geothermal Science, Graduate School of Science Kyoto University, Japan, for providing all facilities for survey and processing data during visiting study. This study was supported by the Technological and Professional Skill Development Program (TPSDP), Physics Department, Universitas Padjadjaran.

Reference

[1] Sudo Y and Hurst AW 1998 Temperature changes at depths to 150 metres near the active crater of Aso Volcano: preliminary analysis of seasonal andvolcanic effects, *J of Volcanology and Geothermal Research* **81** 159–172.

[2] Hase H, HashimotoT, Sakanaka S, KandaW, Tanaka Y 2005 Hydrothermal system beneath Aso volcano as inferred from self-potensial mapping and resistivity structure *J. Volcanology and Geothermal Research* **143** 259–277.

[3] Kagiyama T, Utada H, YamamotoT 1999 Magma ascent beneath Unzen Vulcano, SW Japan, deduced from the electrical resistivity structure *J. Volcanology and Geothermal Research* **89** 35-42

[4] Sharma S P and Kaikkonen P, 1998 Two-dimensional nonlinear inversion of VLF-MT data using simulated annealing, *Geophys.J.Int.*

[5] Kaikkonen P, Sharma S P 1998 Joint Inversion by means of simulated annealing in 2-D:II. The VLF-MT mode and Comparison with the VLF-MT Results, *Phys.Chem.Earth* **23 No 9-10** 901-908.

[6] Tanaka,Y., Eto,T., Nakamura,S., Sudo,Y., Masuda,H,. (1993) : ______, Japanees Version.

[7] Pirttijärvi M, 2006 *Laterally constained two-layer inversion of VLF-MT measurement*, Division of Geophysics Department of Physical Sciences University of Oulu.

IOP Conf. Series: Earth and Environmental Science **29** (2016) 012017 doi:10.1088/1755-1315/29/1/012017

Analysis of Aftershocks Distribution of the June 16, 2010 Yapen Earthquake Based on Stress Changes on the Fault Plane.

I G K S Bunaga[1], I Suardi and N E Deviyanti
State College of Meteorology Climatology and Geophysics, Jakarta, Indonesia

E-mail: satriabunaga@rocketmail.com

On June 16, 2010, an earthquake (Mw 7.0) occurred on the north of the Papua island close to Yapen island. The area of earthquake source has been well identified due to the activity of faults which are formed by collision of four minor plates. The source mechanism of the June 16, 2010 Yapen earthquake was strike-slip fault type with two nodal planes. Determination of the actual fault plane of nodal planes is important for a detailed study of seismotectonics and earthquake risk assessment.In this study, we relocated the mainshock and aftershocks by using the Double-Difference algorithm, and analyzed the stress changes by using the Coulomb stress method on the study area. Furthermore, the determination of the actual fault plane was obtained by using the H-C method (Hypocenter-Centroid method). We used earthquake data from the Meteorology Climatology and Geophysics Agency (BMKG) and the Global Harvard Centroid Moment Tensor (CMT Global Harvard). Study results show that the relocated aftershock distribution lies close to the increase in stress coseismic area. The increase in stress coseismic area occurred at the nodal plane II as the actual fault plane. This results proved that the aftershock distribution was correlated to the actual fault plane.

1. Introduction

Indonesia is one of the earthquake prone countries because of high tectonic activity. There are three major plates and nine minor plates that collided by each other around Indonesian region. These collision zones act as earthquake sources.Papua region isone of the tectonic complex zones in Indonesia. The regionis located among four minor plates namely Sunda, Pacific, Philippine Sea, and Australia plates, respectively (figure 1a) [1]. Due to these tectonic conditions, the occurrence of earthquake is very susceptible. On June 16, 2010 the big earthquake struck Yapen and Waropen districtsin Papua region(see figure1b). Furthermore, the earthquake caused the significant damages and made cracks on the ground trending to direction of 310-350 degree based on BMKG survey.

[1] To whom any correspondence should be addressed.

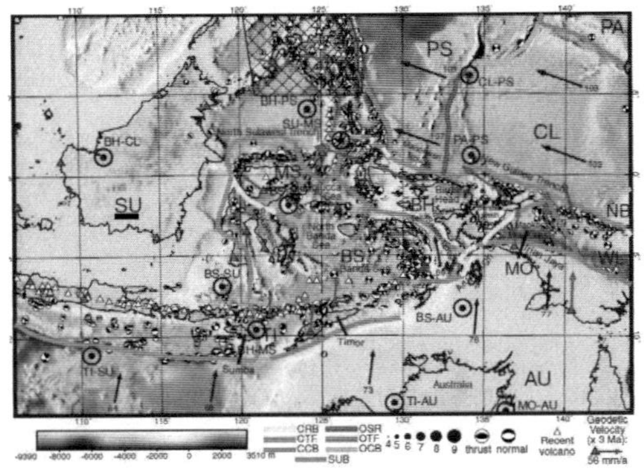

(a)

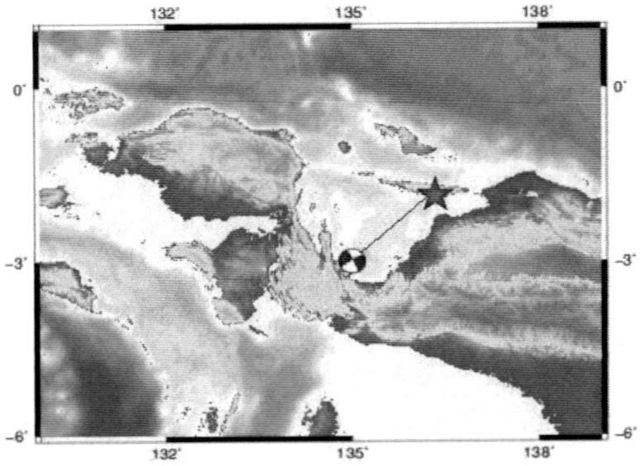

(b)

Figure 1. (a) Regional tectonic features in the Indonesia region, especially Papua region [1]. (b) Red star shows the epicenter of the June 16, 2010 Yapen earthquake.The beach ball shows the focal mechanism from the Harvard CMT (Nodal plane I, strike 241°, dip 82°, slip -10° & nodal plane II, strike 332°, dip80°, slip -172°).

The traces of cracks on the ground represent the direction of the earthquake strike. Knowledge of the earthquake strike can be obtained from aftershocks distribution and coseismic stress in changes [2,3,4,5,6,7]. The aftershocks distribution is also correlated with coseismic stress in changes of the earthquake. The purposes of this study are to determine the actual fault plane based on the patterns of aftershocks distribution and to analyze coseismic stress in changes of the June 16, 2010 Yapen earthquake. Based on these results, we succeeded to improve earthquake hazard assessment in the Papua region.

2. Data

In this study, 98 events were used to relocate the aftershocks M $\geq$ 3 from 13 stations of BMKG broadband seismometer network (figure2). We used the Global CMT data to calculate the stress distribution of coseismic and to determine the actual of fault plane.

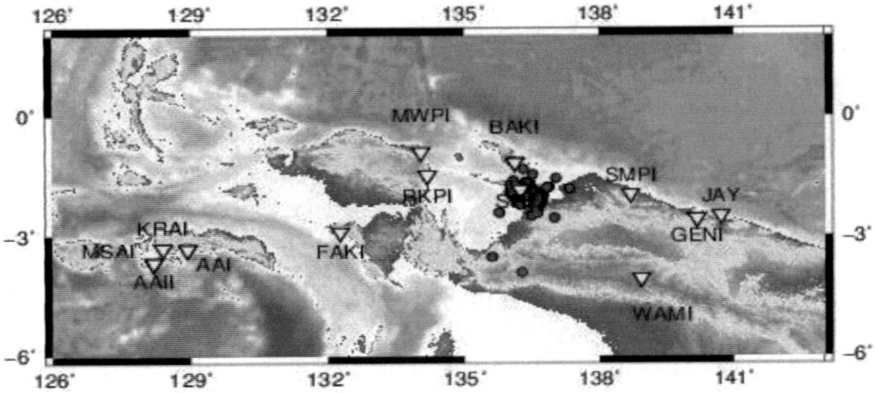

Figure 2. The distribution of 13 stations BMKG broadband seismometer network are markedby yellow triangles.The mainshock and aftershock of the June 16, 2010 Yapen earthquake are marked by red circles.

3. Theory and Methodology

3.1 Double–Difference Algorithm

In this study, we relocated the aftershocksusing Double-Difference algorithm. This method can repair a velocity model as an input in processing, and make the relocation more sense than pre-relocation data[8,9].The basic calculation is based on travel time between two earthquakesdistance which is shorter than the earthquakes-station distance [10].

$$\frac{\partial t_k^i}{\partial m}\Delta m^i - \frac{\partial t_k^j}{\partial m}\Delta m^j = dr_k^{ij} \tag{1}$$

Or rewriten as follows:

$$\frac{\partial t_k^i}{\partial x}\Delta x^i + \frac{\partial t_k^i}{\partial y}\Delta y^i + \frac{\partial t_k^i}{\partial z}\Delta z^i + \Delta\tau^i - \frac{\partial t_k^j}{\partial x}\Delta x^j - \frac{\partial t_k^j}{\partial y}\Delta y^j - \frac{\partial t_k^j}{\partial z}\Delta z^j - \Delta\tau^j = dr_k^{ij} \tag{2}$$

where $\boldsymbol{\Delta m^{ij}}$ is the change of relative hypocenter parameters (Δx^{ij}, Δy^{ij}, Δz^{ij}, and $\Delta\tau^{ij}$) for events i and j. ∂t_k^{ij} is the travel time of earthquake wave (*i* and *j*) to the *k* station. $\boldsymbol{dr_k^{ij}}$ is residual of the observed and calculated events (*i* and *j*) which are recorded by the same station (k). ∂ is the partial derivatives, **m** is the component of the slowness vector. Equation (2) is combined for all of pairs of earthquakes that is solved by linier matrix equation:

$$WGm = Wd \tag{3}$$

with G explains a matrix of size M x 4N (M, number of double-difference observations; N, number of events), *d* is matrix of dr_k^{ij} which its size is Mx 1, *m* is a matrix of model perturbation of Δm with the size 4*N* x 1, and *W* is a diagonal matrix which serves as weight each equation.The algorithm is applied in a packet program namely HypoDD developed by Waldhauser [10].

3.2 Coulomb Stress

We investigated whether the aftershocks are triggered by mainshock by calculating Coulomb stress distribution, ΔCFF. It can be represented as follows [2,3,4].

$$\Delta \mathrm{CFF} = \Delta\tau - \mu\Delta\sigma \tag{4}$$

where ΔCFF is a critical Coulomb failure stress, $\Delta\tau$ is the changes of shear stress related to the slip distribution, μ is an effective coefficient of friction related to pore fluid and normal stress change, and $\Delta\sigma$ is the changes of normal stress, respectively. Coulomb failure stress ΔCFF indicate that a rock will reach a limit to failure[2]. Coulomb failure stress can explain the behavior of stress released due to mainshock, and influence the surrounding area to generate aftershocks. The Coulomb stress changes are displayed into maps by using Coulomb 3 software (USGS) developed by Toda S [11].

3.3 H-C Method

In this study, we identified the fault plane of the June 16, 2010 Yapen earthquake by using the H-C method. The method analyze the geometrical configuration of the Hypocenter (H), Centroid (C), and the moment–tensor solution in Nodal Plane I (NP I) and Nodal Plane II (NP II) [7]. Furthermore, it can investigate the actual fault plane by determining the appropriate location of hypocenter to one of the both nodal planes, and the shortest distance of hypocenter to both nodal planes.

4. Result and Discussion

In this study, we relocated the aftershocks of the June 16, 2010 Yapen earthquake. Most of the aftershocks were located in northwest of mainshock. In lateral, the pattern of relocated aftershocks distribution was almost similar with of pre-relocated one. The pre-relocated pattern in depth showed that the aftershocks were occurred at shallow depth, meanwhile the relocated aftershocks were moved deeper part more than 10 km in depth. (figure 3 and figure 4).The difference of depth between pre-relocated and relocated aftershocks indicated that the relocated hypocenter depth depend on electing the pair of hypocenter in calculating. The elected hypocenter pair should have the shortest distance.

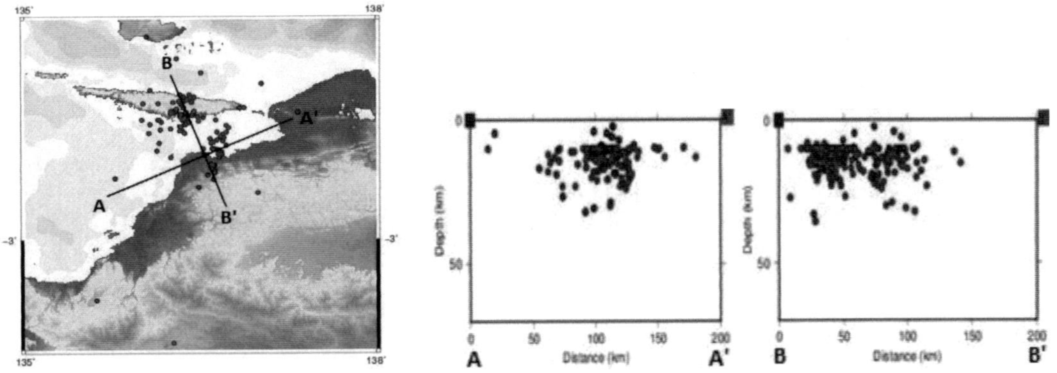

(a) (b)

Figure 3. (a) The initial hypocenter location (BMKG). The line of A-A', and B-B' are cross section lines, respectively. (b) Cross-sections of aftershocks distribution in the depth.

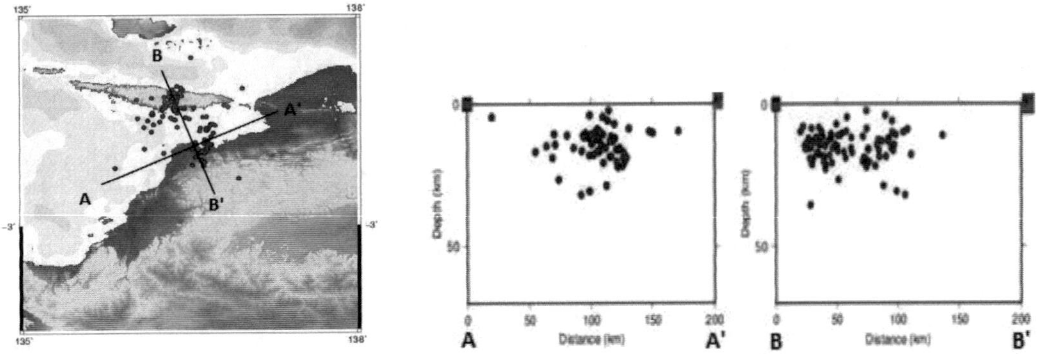

(a) (b)

Figure 4. (a) The hypocenter location relocated by using hypoDD. (b) Cross sections of after-shock distribution in the depth.

We calculated the stress changes of mainshock in order to investigate whetherthe aftershocks were generated by mainshock (figure 5). In Figure 5, the distribution of Coulomb stress shows that there are the increased stress zone at the end of rupture marked by warm color, 0-0.5 bar. This pattern indicates that the aftershocks would be expected to occur in and around this area. It is proven that there are many aftershocks occurred in and around the increased stress zone i.e. in NP II.Meanwhile aftershocks distribution is not represented by the stress in the depth becauseof the uncertainties vertically.We applied the H-C method to analyze the actual fault plane of the June 16, 2010 Yapen earthquake.Figure6shows the position of hypocenter of the June 16, 2010 Yapen earthquake.The solution demonstrated that the hypocenter is closer to the NP II, 13.86 km than to NP I, 38.99 km. So, NP II is the actual fault plane of the June 16, 2010 Yapen earthquake.

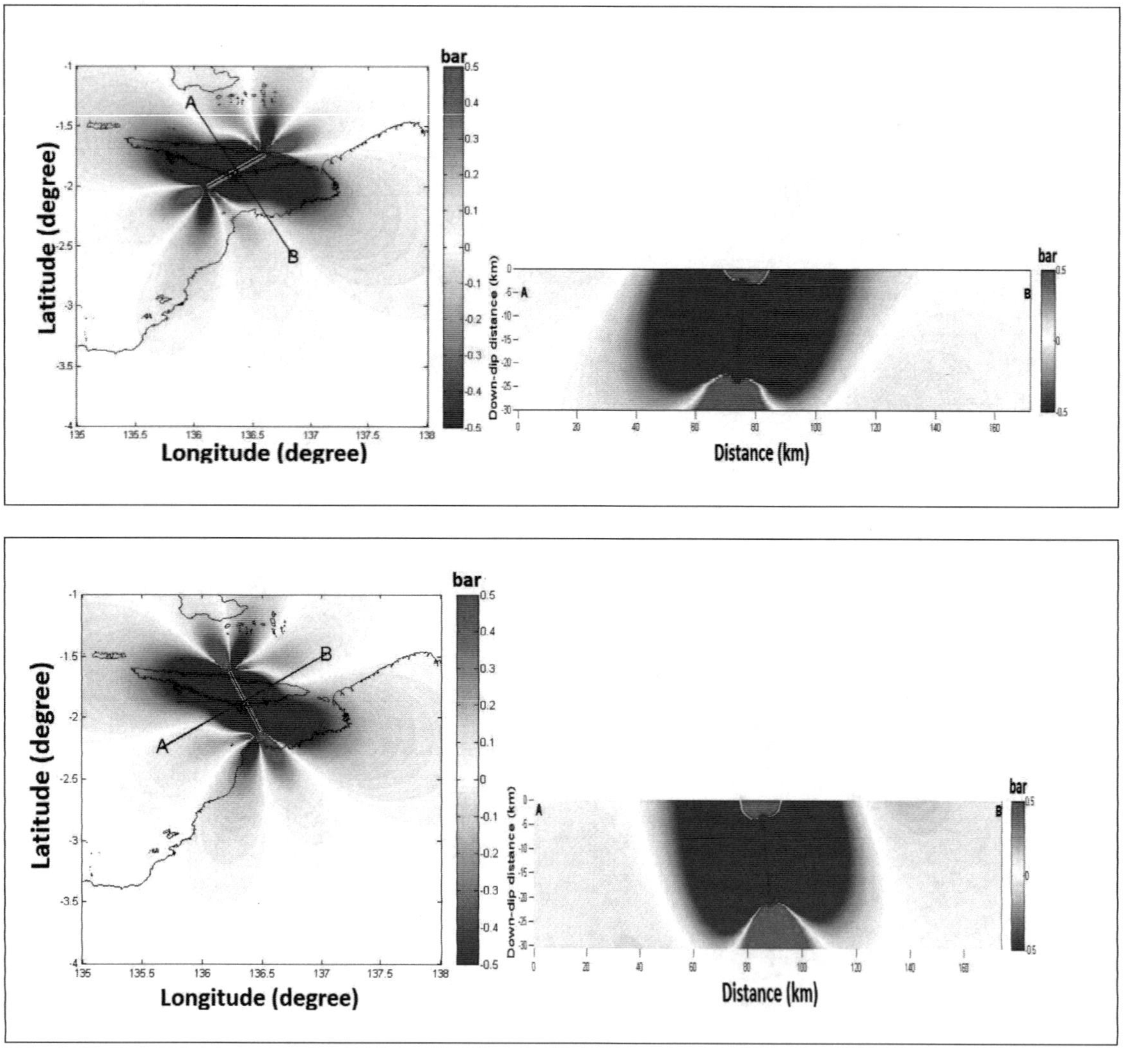

Figure 5. The upper panel shows the Coulomb stress change and vertical cross section A-B (in bars) of nodal plane I and the lower panel represents the nodal plane II.

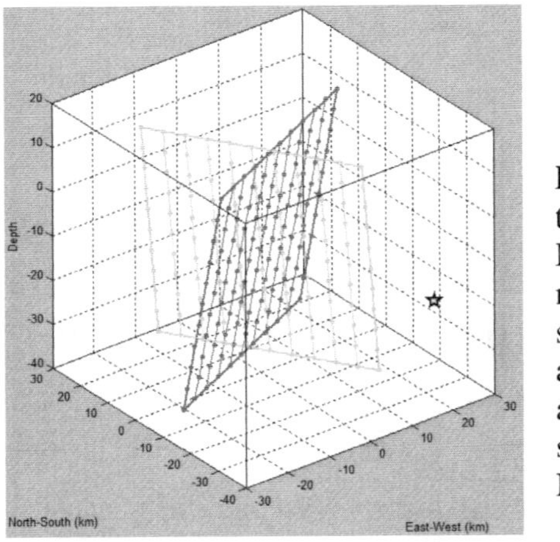

Figure 6. Centroid fault plane is in the middle of the intersection between nodal plane I and II. Mainshock is represented by blue star. NP I is marked by red square. NP II is marked by green square. Because the hypocenter is not located appropriatelyon the one of both nodal plane, the actual fault plane is obtained by determination the shortest distance of hypocenter-nodal plane i.e. NP II (NP I = 38.99 km & NP II = 13.86 km).

5. Conclusion

The relocated aftershock distribution lies close to the increased stress zone. The area of coseismic stress increase was occured in the NP II as the actual fault plane of the June 16, 2010 Yapen earthquake. Moreover, the H-C method shows that the strike direction of actual fault plane is agreement with field observation.

References

[1] Bird P 2003 *Geochem. Geophys. Geosyst.***4** 3
[2] King G C P, Stein R S and Lin J 1994 *Bull. Seism. Soc. Am.* 84 3
[3] Lin J and Stein R S 2004 *J. Geophys. Res.*109 B02303
[4] Raju P S, Gahalaut V K and Kumar M R *J. Geodym.* 46 63
[5] Shis M H, Huang B S, Zhu L, Yen H Y, Chang T M, Huang W G and Wang C Y 2014 *Terr. Atmos. Ocean. Sci*, 255
[6] Yang H, Zhu L and Chu R 2009 *Bull. Seimol. Soc. Am.* 996
[7] Zahradnik J, Gallovic F, Sokos E, Serpetsidaki A and Tselentis A 2008 *Greece. Seis. Res. Let.* 795
[8] Balfour NJ, Cassidy JF and Dosso SE 2012 *Bull. Seismo. Soc. Am.* 90 102
[9] Yang Z X, Waldhauser F, Chen YT and Richards P G 2005 *J. Sei.* 9 2 pp 241-264
[10] Waldhauser F and Ellsworth W 2000 *Bull. Seismol. Soc. Am.* 90 6
[11] Toda S, Stein R S, Dinger K R and Bozkurt B 2005 *J. Geophys. Res.* 110 B05S16
[12] Wessel P and Smith W H F 1991 *Trans. Am. Geophys. Un.*, 72(41), 441ff.

IOP Conf. Series: Earth and Environmental Science **29** (2016) 012018 doi:10.1088/1755-1315/29/1/012018

Valuation of Building strength against earthquake Vibrations using Microtremor Analysis (case study: the city of Surabaya)

Bahri, A.S[11], Utama, W[1], Aini D.N[2], Lutfie,M.N[2]
[1]Deparment of Geophysics Engineering, Planning and Civil Engineering Faculty, Sepuluh November Institute of Technology, Surabaya 60111, Indonesia
[2]Deparment of Physics, Natural Science dan Mathematic Faculty, Sepuluh November Institute of Technology, Surabaya 60111, Indonesia

E-mail : syaeful_b@geofisika.its.ac.id

Abstract. The potential damage due to earthquakes is not only identified by the magnitude and the epicenter distance to a particular region, but also by the topography and it's near surface geological condition. In the inhabited area, such as the urban locations, it is necessary to assess buildings against the earthquake vibration. For this purpose the microtremor vibration analysis was conducted in the city of Surabaya, in order to determine the earthquake-prone buildings and their vulnerability against the vibrations induced by earthquake. The possibility of resonance structure-soil in this specific area was also studied. The microtremor on grounds was analyzed using the Horizontal to Vertical Spectral Ratio (HVSR) and the microtremor inside the buildings was analyzed using the spectrum analysis method, the Floor Spectral Ratio (FSR) and the Random Decrement Method (RDM). This study was performed on ten-borne buildings throughout the region, which represent the one-story buildings, governmental buildings, historic buildings and public buildings. We found that from these buildings, seven of them have low-level, two have low- to medium-level of resonance, and one has a high-level of resonance. The minimum building vulnerability index in the horizontal direction EW (East-West) is 4.55 and in the NS direction (North-South) is 4.14. The maximum vulnerability index was found in the horizontal direction EW (East-West) and the NS directions, which amounts to 143.47 and 171.233.

1. Introduction

Topographically, Surabaya is located in the northern part of the Java Island, which is dominated by lowlands. The majority (80%) of the region is alluvial deposit that spreads along the Porong and Surabaya River. The rest is low rolling hills formed by weathering of tertiary rocks [1]. According to [2], regions formed by alluvial sediments, sandstone, tuff and clay are having a great potential to ground vibration, due to the amplification mechanism induced by the ground and its interaction with the soil-structure during the earthquake. The soft sedimentary rocks are able to strengthen the vibration [3]. Therefore, the average earthquakes damage is more severe compared to the areas with the near surface formed by hard rocks. This phenomenon is proven, for example, in the devastating earthquake of Mexico, in 1985. Although the distance to the epicenter of Mexico City is no less than 390 Km, the induced damage was very severe due to the amplification of horizontal wave acceleration

[1] To whom any correspondence should be addressed.

of Mexico City, due to the fact that the city was formed by soft soil-deposits from compacted bogs having a shear wave velocity of 40 - 90 m/s [1, 3].

This fact clarify that the geophysical micro-zonation acts is required and need to be conducted in to estimate the strength of buildings, using the microtremor analysis. We concentrate our study in Surabaya, in order to classify the earthquake-prone buildings by determining the natural frequency, specifying the vulnerability index of buildings against earthquake vibrations and identifying the possible soil-structures resonance.

2. Theory

2.1 The geology and Seismicity of Surabaya

Figure 1. Geological map of Surabaya taken with permission from [4] with the physiography of mid-Kendeng hills, mid Rembang-Madura hills, north Java alluvium, mid East Java and east of Randublatung curve. The stratigraphic order exposed consists of alluvium deposits (Qa) Kabuh formation (Qpk), Pucangan Formation (Qtp) and the Lidah formation (Tpl).

Judging from the condition of its seismicity, Surabaya is crossed by two active faults that can trigger the occurrence of earthquakes in Surabaya and its surrounding area, namely the fault in the subduction zones along the South of Java with a distance of about 300 Km South of Surabaya and Lasem strike slip fault which is located about 75 Km Northwest of Surabaya. In addition, there is a passive fault located around Surabaya, Watukosek fault that extends from Mount Penanggungan through Sidoarjo Mud Volcano to Gunung Anyar to Madura [5]. Surabaya is located in the class 5 seismicity zone with the PGA values (Peak Ground Acceleration) of 0.15 - 0.2 g, so that Surabaya has is potentially vulnerable to earthquake hazard [6].

2.2 Local site effect from Earthquake

The local site effect, i.e. the local geological and topographic conditions, is one of the causes of destructive earthquakes [7, 8, and 9]. In [1] explain that relationship of the intensity of the earthquake damage to a region are affected by the quake epicenter, distance scale, the size of the earthquake fault zone, the energy released rocks, geological type between the source and the local geological conditions. Differences in local conditions in a region with other regions due to the variation of geological formations, the thickness and physical properties of rocks and soils, surface water and depths of bedrocks, and also underground structures above the region [10].

2.3 Microtremor

Microtremor is a quake vibrations that propagate with the amplitude of the order of micrometers brought about by natural events (e.g. wind and ocean waves) or artificial (e.g., industrial machinery and vehicles) [11]. This method was introduced by [12] and is applied by [2] with the proposed method HVSR (Horizontal to Vertical Spectral Ratio) to estimate natural frequencies and amplification of the local geology of the microtremor data. Later this evolved in the way that the method is able to estimate ground and building vulnerability indices, interaction of soil- structure, and damping ratio of building. In this study, the HVSR method is used to data analysis of microtremor measured in ground and spectral methods, RDM, and FSR for the microtremor data measured in the building. According to [13], in a microtremor analysis of a building, HVSR method is not recommended to despite the natural frequency of its estimation result is logical. This is caused by there is no basic theory in its application so that it cannot be assumed that vertical and horizontal spectrum on the underground levelis the same.

3. Method

3.1. Analysis of microtremor data measured in Ground

This research uses a set of portable microtremor instrument consists of three component digital seismograph station with short period manufactured by Taurus (Canada) with short-period seismograph feedback sensor type DS-4A and equipped digitizer (Data Logger). Microtremor measurement of the soil conducted on 10 point location, adjacent to the building that will be analyzed.

Geopsy software is use for the data processing. Each vertical and horizontal components of the spectrum is divided into 20-40 S long window tapper with 5% cosine function. Furthermore, the analysis of Fourier spectral of each window will change the initial microtremor data from time domain into the frequency domain. Smoothing process of FFT results implement [9] filter, with bandwidth coefficient of 40.

Smoothed data is analyzed to use amplitude spectrum ratio (H/V spectrum ratio) method of the two amplitude of horizontal and one vertical Fourier spectral being estimated using Equation (1) [14]:

$$R(T) = \frac{\sqrt{F_{NS}(T)^2 + F_{EW}(T)^2}}{F_Z(T)} \tag{1}$$

where $R(T)$ = horizontal to vertical ($H/$V) ratio spectral, F_{NS}, F_{EW} and F_Z *are* Fourier spectral *in NS*, *EW* and Z (vertical), respectively.

Furthermore, spectrum is plotted as the H/V spectral for each measurement point. Typically, the frequency of peak of H/V spectrum is the natural frequency of the local site effect. Therefore, estimation of vulnerability index of ground can be determined from natural frequency and amplification.

3.2. Analysis of Microtremor measured in building

Microtremor Data taken on 10 buildings spread on the whole area of Surabaya (table 1) with the criteria of public buildings, Government buildings and historic buildings. Parallel with the above activities, also taken soil data that represents the subsurface Geologic conditions of each building.

Microtremor measured in building recorded along 15 minutes each floor of the building, and 20 minutes for the ground using the same process of measurements. Analysis of microtremor data measured on building used three methods, namely; spectral analysis, RDM and FSR. Spectral analysis is done by dividing the window spectrum and continued with FFT (Fast Fourier Transform) analysis. Because the result of the FFT analysis is random, smoothing technique [7] is used with a bandwidth of 40. Calculation of average spectrum of the entire window is then applied. The natural frequency is identified as the frequency with largest peak amplitude.

To find out the possible soil-structures resonance underneath when the earthquake occurred, the calculation is done based on the following equation:

$$R = \left| \frac{fb - ft}{ft} \right| \times 100\% \quad (2)$$

According [13] classify the vulnerability level of soil-structures resonances against earthquake in three criteria, i.e. high vulnerability to the difference frequency ±15%, medium vulnerability for 15-25% values and low vulnerability for higher than 25%.

4. Results and Discussion

The natural frequency in the ground shows geological structure in the local area. The magnitude of natural frequency is inversely proportional to the depth of bedrocks and is directly proportional to the velocity or the average density of the subsurface structure of the local.

Tabel 1. Analysis result of ground and building microtremor and resulted ground natural frequency, building natural frequency and building resonance to the ground (R: Low and T: High)

Code	Building	Ground	Ground F (Hz)	Damping Ratio		Building Spectrum		Building RDM		Building FSR		Resonance
				EW	NS	EW	NS	EW	NS	EW	NS	
M97	Lt.7 Gedung Keuangan Negara	T22	1.1756	7.79	5.96	1.18586	1.20058	1.32	1.19	1.21549	1.18586	T
M92	Lt.2 Gedung Keuangan Negara	T22	1.1756	7.03	6.1	1.23058	1.12876	1.2	1.22	1.27698	1.10125	T
M114	Lt.4 PDAM Surabaya	T32	1.2894	6.73	7.85	2.95482	2.95482	2.95	2.88	2.95482	2.91859	R
M123	Lt.3 IntilandTower	T33	2.3151	4.29	1.37	1.03537	1.15696	1.07	1.15	1.03537	1.12876	R
M202	Lt.2 Gelora Bung Tomo	T15	0.7072	4.6	6.34	1.44465	1.32512	1.5	1.33	1.37509	1.35823	R

Tabel 2. Estimation of building and ground vulnerability indices and building maximum acceleration

Code	Ground	Ground Accl	Building F		Building A		H	Average Kb		Max Acceleration	
			EW	NS	EW	NS		EW	NS	EW	NS
M97	T22	3.56	1.24	1.19	2.58	3.87	21	20.18	32.86	347.70	213.49
M92	T22	3.56	1.24	1.15	1.40	1.77	6	38.79	56.47	180.85	124.22
M114	T32	3.08	2.95	2.92	2.55	2.269	16	4.62	4.22	1760.29	1928.22
M123	T33	2.57	1.05	1.15	5.55	9.32	12	106.89	149.84	91.06	64.96
M202	T15	5.76	1.44	1.34	0.29	0.23	8	4.55	4.14	954.77	1049.63

The application of HVSR method in the Surabaya area shows the natural frequency of te 15 point of measurement in Surabaya which were used to estimate the possibility soil-structures resonances, vulnerability index of ground and vulnerability index of building. The value of soil frequency ranges between 0.7072 - 2.3511 Hz. The research results showed that the natural frequency of local site in Surabaya is small (table 1), which means that the soft layer is thick enough in Surabaya. This was confirmed by [4], that Surabaya is generally dominated by lowlands, 80% of the area is alluvial deposits and the rest is low rolling hills formed by soil weathering results of tertiary rocks/old.

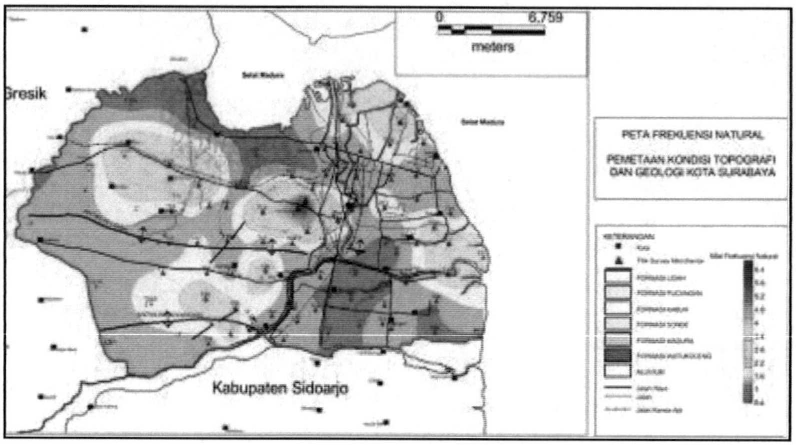

Figure 2. Map of that the natural frequency of local sites Surabaya is low which is marked by purple – red colour on the figure with an average frequency of 0.5 - 2.7 Hz

Table 1 shows the value of building frequency using the spectral analysis, RDM and FSR approaches. The analysis of each building resulting horizontal natural frequency value (NS and EW). The natural frequency values obtained at the Graha Pena is varied. Research on the building was conducted on several floors to prove whether the natural frequencies of each floor in a building are the same. Therefore, it requires only a single measurement at one floor.

To support the result of natural frequency data, damping ratio analysis, vulnerability index and maximum acceleration can be held down the building needs to be calculated. The vulnerability index and the maximum acceleration of the building can withstand is estimated from FSR results analysis. Also in the case of natural frequency analysis results, the parameter index of the vulnerability of buildings and the maximum acceleration that the buildings can withstand also pointed out that the strongest building is PDAM Surabaya, while the weakest is Intiland Tower building.

Estimation of damping ratio is different with natural frequency, vulnerability index and the maximum acceleration that is capable of being held by buildings. Damping ratio estimation results show that Intiland Tower has the lowest strength while STIKOM Surabaya has the weakest buildings strength. But the strength difference is not noticeable. This is possible because the method used is based on the single degree of freedom (SDOF) whereas in practice building naturally has multi degree of freedom (MDOF). Next in the classification of strength building determination, this parameter can be ignored.

5. Conclusion

The potential damage to buildings as a result of the vibrations of an earthquake can be estimated using natural frequency parameter, the possibility soil-structures resonance, and the maximum acceleration of the vulnerability index which the building can withstand. From the analysis results can be shown that Rusun Penjaringan Sari, Antariksa Hotel, Tujuh Belas Agustus University, Sukolilo Dian Regency, East-Jawa Police Department, Ciputra Apartment, EAST JAVA Senate, PDAM Surabaya, Intiland Tower, Siola Tunjungan City, Bhayangkara University, UNESA Physics Department, Firehouse Department, Bung Karno Stadium has the potential resonance between the structure and soils. The Intiland Tower and the STIKOM building has the weakest strength, while the most powerful buildings are Surabaya Firehouse Department, National Finance building and Surabaya PDAM building and the others have medium strength.

References

[1] Towhata I 2008 *Geotechnical Earthquake Engineering Springer* SBN: 978-3-540-35782-7 Japan

[2] Nakamura Y 1989 A method for dynamic characteristics estimation of subsurface using microtremor on the ground surface *Quarterly Report of Railway Technical Research Inst. (RTRI)* **30**, p 25–33.

[3] Tuladhar R 2002 *Seismic Microzonation of Greater Bangkok Using Microtremor Observations* Thesis Asian Institute of Technology School of Civil Engineering Thailand

[4] Sawolo et al 2009 The Lusi mud volcano controversy: Was it caused by drilling? *Elsevier Marine and Petroleum Geology* **27** 1651–1657

[5] Sukardi 1992 *Peta Geologi Lembar Surabaya dan Sapulu, Jawa* Pusat Penelitian dan Pengembangan Geologi (PPPG) Bandung

[6] Irsyam M et al 2010 Development of Seismic Hazard and Risk Maps for New Seismic Building and Infrastucture Codes in Indonesia, *Proceeding the 6th Civil Engineering Conference in Asia Region: Embracing the Future through Sustainability* ISBN 978-602-8605-08-3

[7] Nakamura Y 2000 Clear identification of fundamental idea of Nakamura's technique and its applications *Proc. XII World Conf. Earthquake Engineering* New Zealand 2656

[8] Herak Marijannet al 2009 HVSR of Ambient Noise in Ston (Croatia): Comparison with Theoretical Spectra and with The Damage Distribution after The 1996 Ston-Slano Earthquake *Springer Science+Business Media B.V.*

[9] Warnana D D Triwulan Sungkono Utama W 2011 Assessment to the Soil-Structure Resonance Using Microtremor Analysis on Pare-East Java, Indonesia *Asian Transactions on Engineering* (ATE ISSN: 2221-4267) **Volume 01** Issue 04

[10] Oliveira S C 2006,Assessing and Managing Earthquake Risk *Springer*

[11] Tokimatsu K 1995 Geotechnical site characterization using surface waves *In Proc. 1st Intl. Conf. Earthquake Geotechnical Engineering* Ishihara (ed) Balkema 1333-1368

[12] Kanai K and Tanaka T 1961 *On Microtremor VIII, Bull.Earthq. Res.Inst.*Tokyo University **Vol.39** pp 97-114.

[13] Gosar A 2010 Site effects and soil-structure resonance study in the Kobarid basin (NW Slovenia) using microtremors *Nat. Hazards Earth Syst. Sci.* **10** 761–772

[14] Konno K and T Ohmachi 1998 Ground-motion characteristics estimated from spectral ratio between horizontal and vertical components of microtremor *Bull. Seism. Soc. Am.***88** no. 1 228–241.

Fractal Characteristics of Geomorphology Units as Bouguer Anomaly Manifestations in Bumiayu, Central Java, Indonesia

Andi Agus Nur[1], Ildrem Syafri[1], Dicky Muslim[1], Febri Hirnawan[1], Pradnya P Raditya[2], Murni Sulastri[2], Fikri Abdulah[2]
[1] Postgraduate Program, Faculty of Geological Engineering, Padjadjaran University, Jl. Raya Jatinangor 21st Km, Jatinangor, Indonesia
[2] Geomorphology and Remote Sensing Laboratory, Faculty of Geological Engineering, Padjadjaran University, Jl. Raya Jatinangor 21st Km, Jatinangor, Indonesia

E-mail: noor_geo79@yahoo.co.id

Abstract. Bumiayu in Central Java, Indonesia, has a typical landform characteristics. Differences of topography on each geomorphological unit indicated by the value of fractal dimension. This research provides important information on the influence of geomorphology conditions and subsurface geological phenomenon of research area based on fractal application. This research methodology relies on laboratory analysis and field observation. Landform is a characteristics of Bouguer anomaly contour manifestation. It is indicated by occurences of significant correlation between the Bouguer anomaly countour and geological cross section as well topography contour slope and Bouguer anomaly contour slope. Based on spatial analysis, morphology of research area is dominated by very high steep hills (more than 60%). Result of Bouguer anomaly countour analysis also shows that research area dominated by very high steep hills (more than 55%). Statistical analysis between the fractal value of lineament in *Digital Elevation Model* and fractal value of lineament in Bouguer anomaly countour as well the fractal value of topography countour and fractal value of Bouguer anomaly countour shows that the relationship was not significant. Further, the entire result of this verified research shows clearly that geomorphology conditions represents subsurface geological phenomenon.

1. Introduction

Bumiayu has a complex geological structure and lithology types including plain to steep hills morphology. Variations of this landform are affected by endogenous factors (tectonic style) as well as exogenous factors (weathering and erosion). In understanding the phenomenon of land formation, especially in terms of the influence of endogenous factors, required a special instrument that is able to represent visually Bouguer anomaly phenomenon. One method that can be used is a method of Bouguer anomaly. Qualitative analysis of Bouguer anomaly contour has a tendency to weakness in the form of perceived differences in interpretation of data. Simplification method Bouguer anomaly contour visualization into a simpler form is very important to do. One of the instruments to analyze the quantitative interpretation and simplify the contour map is fractal method. However, the method of fractal has not been developed comprehensively, in particular when it is used to identify the relationship between the landscape and the Bouguer anomaly phenomenon. Based on the presentation, the authors are interested in knowing how the relationship remedies landscapes and Bouguer anomaly phenomenon and to examine various factors that influence such as lithology and geological structure.

2. Methodology

The process of the formation of structure deformation illustrates the long history of an area. Indications of structures that can be recorded on the Earth's surface can be analyzed using several methods, including: Processing of lineament using DEM, performed on the image of ASTER GDEM Bumiayu area. The ASTER GDEM image is processed using the program map information for later withdrawal of lineament, including lineament ridges and lineament of the valley with the help of the Global Mapper program.

Gravity anomaly map is a map that overlay Bouger Anomaly countour and regional Geological map of the study area [3][6] using simple gravity anomaly Bouguer density of 2.18 g / cm^3. Each observation point anomaly derived from equations:

$$AB = g_{obs} - g_t + FAC - BC + TC \quad (1)$$

Bouger Anomaly (AB), using Equation (1) is total of Gravity Observation (g_{obs}), Theoritic gravity (g_t), Free Air Correction (FAC = 0,3086 mGal/m), Bouguer Correction slab (BC = $2\pi G\rho$ mGal/m), Topography Correction (TC). Bouguer anomaly map (Figure 2.1) derived from about 400 observation points Bouguer anomaly that covers the whole of the study, by observation point distribution randomly (random).

The methods used in the calculation of the fractal dimension using a box counting method or the dimensions of the box. This method can be applied to objects that are self-similar statistical or statistical self-affine fractal. The box counting method is done by making certain sided grid (r) on a fractal object. [1][2][4][7][8]

We denote the cut or the set-fractal as F and its sum inside the surrounding r-sided box is Nr (F). The calculation of this value is done iteratively with until the smallest value of fractal is achieved. In this way, the small variation of F value to r can be plotted in a logarithmic graph. The dimension of this calculation can be determined from the slope of the plot.

$$D = \lim_{r \to 0} \frac{\log Nr(F)}{-\log r} \quad (2)$$

The same way done well to get the value of the fractal dimension (equation 2) for other variables, and then tested using statistical probabilistic.

3. Result and Disscusion

3.1 Surface Data

Folding direction (either Syncline or anticline) generally has a direction Northwest-Southeast and West-East. Folding form of Syncline in the area of research are Syncline Cisaat and Penujah. Based on observations of the Digital Elevation Model (DEM) and field data [5], an indication of the geological structure in the form of transform dextral fault can be analyzed based on the evidence in the form of offset, the lineament of the ridge, as well as data fractal.

The landscape is a product of tectonic processes and the formation lithology. The landscape in the study area is dominated by fine clastic lithology types. The landscape of steep and very steep hills located in the northwest, South, Northeast-Southeast with different characteristic lithology.

3.2 Subsurface Data

Based on data from the overlay of Bouguer anomaly contour map and topographic contour can be interpreted that in the Central part of the study area are dense anomaly contour, while the topographic contours visible tenuous. This is because the condition of the surface has experienced high attrition with an indication of the form of fine clastic lithology type. In the Northeast to the Southeast Bouguer anomaly contour seemingly tenuous, this is because in these areas there is a region of the magma chamber, while the surface region is an indication of coarse clastic lithologic type. From the results of spatial analysis between the two maps can be analyzed that there is a difference between the surface and geologic conditions below the surface phenomena. This can be caused by several factors, including tectonic activity, erosion intensive, sedimentation, and lithification rocks. While the phenomenon is quite different under the surface because the subsurface region is still plagued Oleg erosion processes and activities such as rock lithification

Based on the analysis and interpretation of landforms Bouguer anomaly contour, in the Northeast part of the study area are likely to experience the eruption of magma to the surface caused by the eruption of Mount Slamet. This led to the formation of a landscape seen in the form of plateaus or hills. In the Central and Southern regions experienced research indicated a fairly intensive process of erosion. This was shown by the data morphometry Bouguer anomaly contour that shows landform steep hills, while the shape of the landscape seen today by exposure to map morphometry or slope is very gentle form of plain land.

3.3 Fractal Data

Statistical interpretation of results between fractal lineament Bouguer anomaly contours and lineament fractal surface using statistical analysis test t-Test. Based on statistical analysis test, t-test showed that the value of t = -7.1136, and table = 2.0032. T-count table may mean that statistical analysis t-test were not significantly different, and Ho is unacceptable, so the value of fractal lineament of Bouguer anomaly with a value of fractal lineament of maps Digital Elevation Model (FLDEM) is the same, or not significantly different.

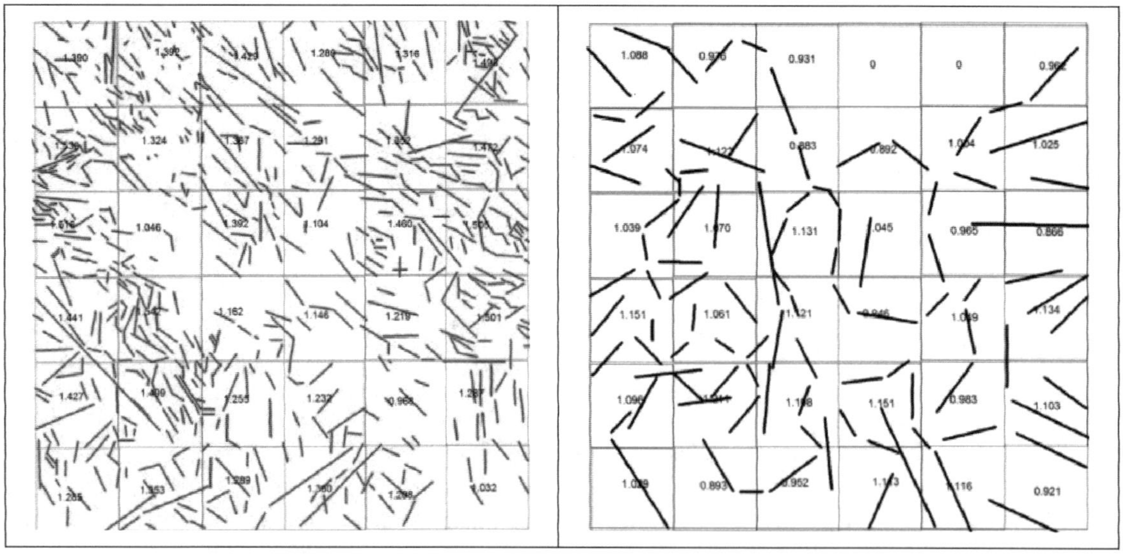

Figure 1 Fractal Lineament of surface and subsurface

Based on the results of statistical test, fractal Bouguer anomaly contour and fractal topographic contours, table value obtained from the calculation of the statistical analysis of -7.7423 and 2.0129 T-count (Figure 1). Based on this data, Ho is accepted that there is a relationship between the Bouguer anomaly contour topographic contours. H1: $\mu 1 \neq \mu 2$ there is no real difference between the value of fractal contour Bouguer anomaly (FCBA) with fractal topographic contours of topographic maps (FCT).

The results of different test t-test Variance Two Samples Assuming Unequal show t-count <t-table. It can be interpreted that there is no significant difference between the value of the fractal Bouguer anomaly contour gradient (FCBA) with topographic slope value of the topographic map (FCT) with significance level of $\alpha = 5\%$. Therefore the surface contour (topography) is different under the surface contour (Bouguer anomaly contour).

Different test results Two Samples t-test showed t-count Assuming Equal Variance <t-table. It can be interpreted that there is no significant difference between the value of Bouguer anomaly contour gradient (FCBA) with topographic contour value (FCT) with significance level $\alpha = 5\%$. Statistical test results using t-Test can be seen that the value of the Bouguer anomaly contour topographic contour values were not significantly different.

Below the surface contour basically has not undergone sedimentation or erosion, while the contour of the surface has undergone sedimentation in a long time. In addition contours on the surface have also experienced the effect of erosion and deposition of volcanic material.

Data values drainage density (Dd) and value drainage fractal pattern that has been obtained from the previous analysis will be tested hypotheses regarding the correlation between the density of the existing drainage pattern of the fractal pattern value stream drainage patterns in the area of research (Figure 2). This hypothesis test is also intended to determine that the value of fractal patterns jetting flow patterns play a role on the drainage pattern density value (Dd) in the research area.

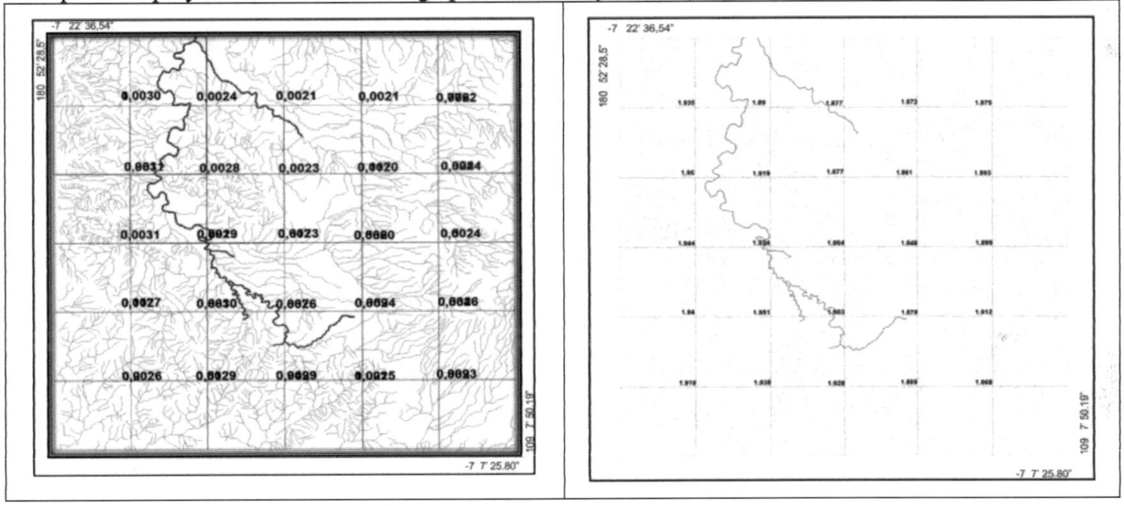

Figure 2 Drainage Density (Dd) of Research Area and River Fractal

4. Conclusion

The result of this research supports the assumption that the surface geomorphology (landscape) in Bumiayu area and the surrounding is a reflection of the Bouguer anomaly. The application of the fractal concept supports the conclusion that the characteristics of fractal unit geomorphology represent the Bouguer anomaly in this particular area. The T-test statistical analysis on the fractal dimension of the Bouguer anomaly contour and the contour topographic have a relationship that is insignificantly different, while the value of river density (Dd) to the value of fractal rivers, have a high correlation, this shows the value of the fractal dimension of river flow patterns correlated with deformation patterns in the study area.

5. Acknowledgement

Authors are grateful to all those who have helped in the research, preparation until the publication of this research. Primarily to Laboratory of Geomorpholoy and Remote Sensing and Laboratory of Geophysic Faculty of Geology, Padjadjaran University.

References

[1] Aviles, C. A., and C. H. Scholz., *1987 Journal of Geophysical Research* **V.92** no.B1 331-344.
[2] Bunde A. and Havlin S., 1994. *A brief introduction to fractal geometry. In 'Fractals in Science', pp 1-25*, Springer-Verlag, Berlin
[3] Djuri, M., Samodra, H., Amin, T.C., Gafoer, S. 1996. Geological Map of Purwokerto and Tegal Quadrangle, Java. Scale 1 : 100.000. Bandung: Geological Research and Development Centre
[4] Falconer, K., 1990. Fractal Geometry. *John Wiley & Sons*
[5] Hirnawan, F., 2009, *A Measure of Intense tectonic in West and Central Java Through Manifestation of River Basin Morphometry Development on Quaternary Volcanic Deposits,* Jurnal Geologi Indonesia, Vol. 4 No.4 Desember 2009: 285-300
[6] Kastowo, dan Suwarna, N., 1996. *Geological Map of Majenang Quadrangle. Scale 1:100.000.* Bandung: Geological Research and Development Centre
[7] Mandelbrot, B. B., 1982. *The Fractal Geometry of Nature.* San Francisco, W. H. Freeman and Company
[8] Turcotte, D. L., 1992. *Fractals and Chaos in Geology and Geophysics*. Cambrige University Press. 221 p.

Applied velocity versus offset (VVO) to validated & characterized fracturing zone in intra Baturaja Formation, South Sumatera Basin

Hilman Mardiyan, Saifatur Rusli
Samudra Energy, Cyber 2 Tower, 26th floor, Jakarta, 12950, Indonesia

E-mail: hilman.mardiyan@samudraenergy.com

Abstract. The velocity versus offset (VVO) as new geophysical method can be applied to detect some geological phenomenon, such as hydrocarbon trap, structural-fracture anomaly, facies changes, etc. The VVO method is data driven, based on the normal move out equation (NMO) and measuring the local event correlation between adjacent traces to get velocity gradient attributes which is derived from cross-plotting the velocity versus offset (VVO). This paper is describing applied VVO model that controlled by well data which indicated fracture from logs data, especially Resistivity Imager Logs or Formation Micro Imager (FMI). Images FMI logs data at Intra-Baturaja Carbonate Formation (BRF) in South Palembang Sub-basin (SPB), South Sumatera, shows vugs with fractures which orientation is roughly NNW-SSE. Meanwhile, the 2D NMO seismic gathers indicated those all as hockey stick at far offset. By applying VVO method, hockey stick can be identified and then used to validated, characterized and localized where the fracturing zone in intra-Baturaja Formation is. Laterally, VVO quantified as velocity gradient attribute which associated with geological model as the fracturing zone in study area. Characterization fracturing zone in Intra Baturaja Formation as geological lateral model by design is a challenging task for most exploration and production. In term of exploration where limited data is available, it can be used step ahead as carbonate fracture reservoir candidate in proven area and adjacent, especially in SPB South Sumatra.

1. Introduction

The South Sumatra Basin (SSB) Province consists of several structural sub-basins with Tertiary sedimentary section lying unconformably on the eroded and faulted topography of pre-Tertiary metamorphic and igneous rocks [1].

1.1. Regional Geology & Stratigraphy of South Sumatra Basin (SSB)

SSB is one of the most hydrocarbon prolific Tertiary back-arc basins today which located primarily onshore Sumatra, Indonesia. The province covers an area of approximately 117,000 km^2 consists of Tertiary half-graben basins infilled with carbonate and clastic sedimentary rocks. It was formed as a pull-apart basin related to NW-SE trending dextral strike-slip faulting.

SSB is divided into several sub-basins: Jambi, North Palembang, Central Palembang, South Palembang, and Bandar Jaya Basin [2] (figure 1). It experienced three tectonic deformation phases: Mesozoic compressional, Eocene-Oligocene extensional and Pliocene-Pleistocene compressional tectonics.

The general stratigraphy of SSB is one consisting of Pre-Tertiary and Tertiary Rocks. Basement Pre-Tertiary rocks consist of meta-sediment rocks, granitic and ultrabasic igneous rocks, volcanic rocks aged range from Permo-Carboniferous (248-354 Ma) to Mesozoic (Jurassic-Cretaceous, 170-110 Ma). Tertiary Rocks consist of Lemat Formation, Talang Akar Formation, Baturaja Formation, Telisa (Gumai) Formation, Air Benakat (Lower Palembang) Formation, Muara Enim (Middle Palembang) Formation, and Kasai (Upper Palembang) Formation (figure 2).

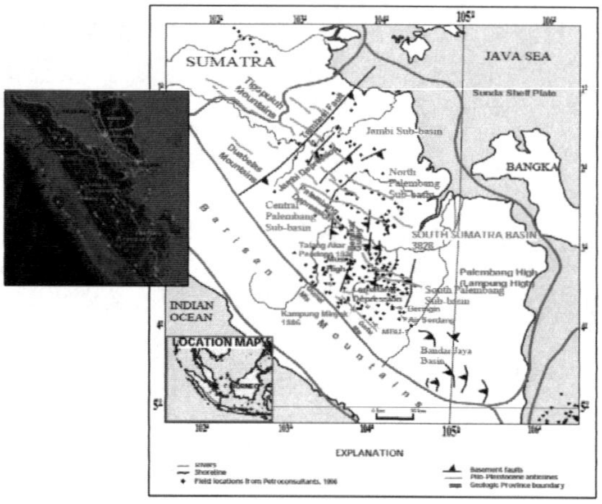

Figure 1. Research area: South Palembang Sub Basin, in proven area and adjacent South Sumatra Province [1].

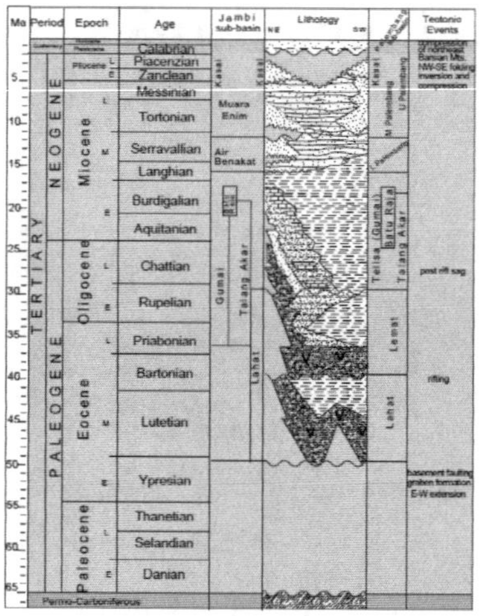

Figure 2. Generalized stratigraphic column for the SSB. ([3] and [4])

1.2. Fracture Identification: References Cited

Fractures are planar features with no apparent displacement in geological formation [5]. Fractures in limestone can be found in a wide range of scales from millimeters to tens of meter long. In general, four different types of fractures can be found: shear fractures, extensional fractures, stylolites and vuggy-a solution enlargement fractures [6]. Fracture-porosity is estimated from core laboratory tests or formation imaging using FMI or FMS methods, because there are no conventional well log tools that can directly determine fracture porosity [2]. Resistivity images respond to electrical contrast as fractures are generally resistive or conductive relative to the host rock. (Figure 3).

Some fractures study was conducted specially in Baturaja Formation SSB. showing fractures in BRF Fm. located on Southern part of SPB-SSB using FMI data proved as the main tools to help the next development well locations and reservoir zones. FMI images showing breakouts and drilling induced fractures in these Well which has NNW-SSE where localized vugular porosity as porous zones (cavities or vugs). On the other hands, Baturaja Formation (BRF)-SSB in thin sections is mouldic-micro vuggy with some fractures.

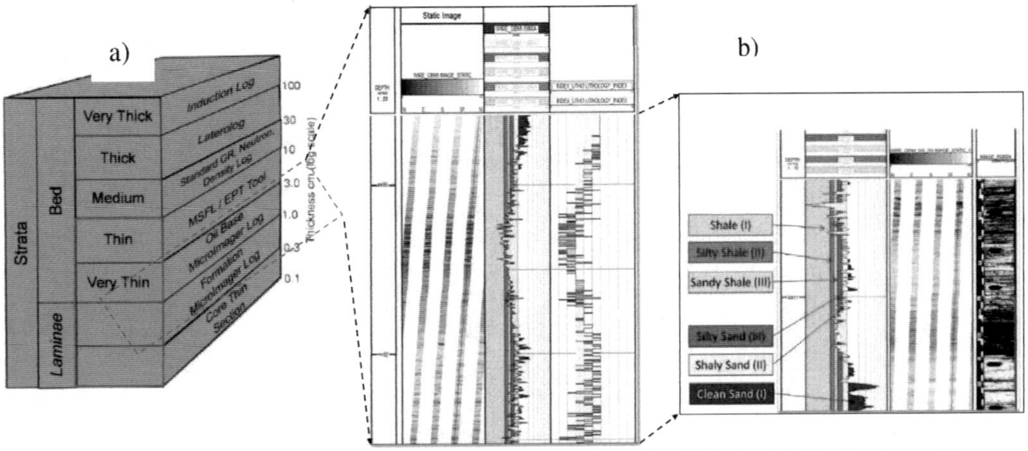

Figure 3. a) Bed classification and tools used for analysis, b) Lithological index log. The green color refers to the shale group and the yellow color refers to the sand and carbonate groups [7].

Much effort has focused on measuring amplitude variations with offset (AVO) for seismic reflections from a fractured reservoir [8, 9, and 10]. Amplitude variation with offset and azimuth (AVOA) has been widely used to identify the orientation of vertical fractures with variable levels of success [11, 9, 12, and 13]. In these applications, the fractures are assumed to be small relative to the wavelength of the seismic waves. This allows the fractured medium to be modeled using an equivalent anisotropic medium.

The presence of hydrocarbons, especially gas in sand reservoirs, commonly creates the 'hockey stick' effect in the common midpoint (CMP) gathers when normal moveout (NMO) corrections have been applied, indicating that a faster NMO velocity is required to flatten the reflection events. Prior to scanning the amplitudes within CMP gathers, it is common practice to apply a time-offset mute to remove overcorrected traces, or to apply trim statics to force reflection events to be flat. A new method for calculating the correct *Vrms* automatically was developed without scanning the set of velocities, but by considering only the Local Even Correlation (LEC) of adjacent traces from the nearest to far end offset. The proposed method honors changes of velocity in each offset. Therefore, any anisotropic characters in the CMP gather will be measured automatically. This principle is simple, but requires high signal-to-noise-ratio data and good event continuity along the offsets within the CMP gather.

2. Velocity versus Offset

2.1. Offset-dependent velocity estimation

Velocity extraction from seismic data is formulated from the NMO equation. The CMP gather is first NMO-corrected using the initial stacking velocity, V_{Stk} [14]. An offset-dependent moveout velocity, Vh_j, is then calculated independently for each offset X_j by assuming moveout is hyperbolic:

$$\frac{1}{Vh_j^2} = \frac{(\Delta t_j + T_j)^2 - T_0^2}{X_j^2} \quad (1) \qquad \text{Where } T_j : \; T_j = \left(T_0^2 + \left[\frac{X_j}{V_{Stk}} \right]^2 \right)^{1/2}$$

T_j is the traveltime at offset X_j assuming hyperbolic moveout at the initial stacking velocity.
In Equation 1, Δt_j is the residual moveout at offset Xj, and is equal to the time shift measured using LEC method. The nearest-offset trace is used as the reference for measuring residual moveout on the other traces. *Vh* can be calculated for each individual offset across the CMP gather. Equation 1 thus acts as an operator to transform CMP data from amplitude gathers into velocity gathers. The VVO analysis is then carried out on the velocity gathers. The output *Vh* gather can be stacked or averaged to

produce a high resolution velocity field for NMO correction. Partial stacking can also be done to extract near- and far-offset velocity fields. The difference between far- and near-offset *Vh* values provides a quick scan of velocity gradient in the VVO method.

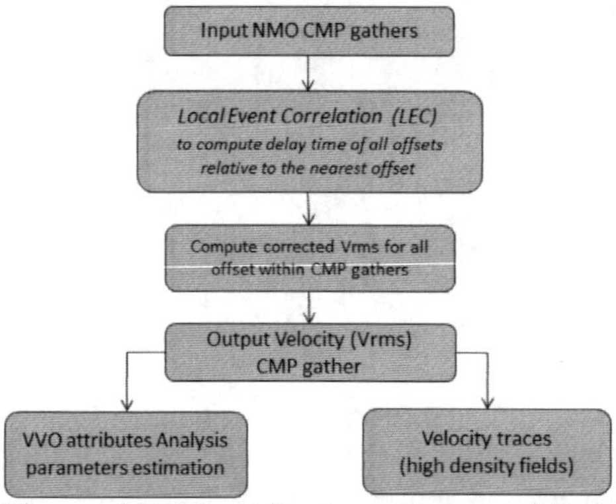

Figure 4.VVO attribute analysis workflow

The velocity analysis window was set to 40 ms and designed to follow the reflection curvature across offsets. The time step for calculating *Vh* is half of the analysis window, so the window steps downwards with 50% overlap over the record length. The complete workflow for VVO attribute analysis is shown in Figure 4 above.

3. Fracture Modelling

A great deal of research has focused on the applicability of finite difference (FD) methods to modeling seismic scattering phenomena. In particular, [15] test the performance of [16] equivalent medium anisotropic cell approach for FD modeling of discrete fractures. The FD method is advantageous when modeling the scattering effect of discrete fractures because of its stability over a wide range of material property contrasts and its ability to model all wave types with minimal numerical dispersion and anisotropy.

Our numerical experiments use a simple reservoir geometry consisting of three horizontal layers. The first and third layers bound the reservoir and are homogeneous and isotropic with the same material properties. The model uses a standard surface seismic reflection acquisition geometry, however, the source and receivers are embedded within the first layer to eliminate free surface effects. The receiver spacing is 20-m with the maximum source-receiver offset at 4000-m which corresponds to a maximum angle of incidence of 45° relative to the top of the reservoir (Figure 5).

Snapshots of the wavefield in the presence of discrete fracture zones highlight the different scattering events and their interactions. Figure 6 shows the vertical component of velocity after 0.6-second. We see that the P-wave reflection from the top of the reservoir remains relatively coherent, while the P-wave reflection from the base of the reservoir is much more effected by the presence of the discrete fractures. This figure also illustrates the difference in coherence of the scattered wavefield normal and parallel to the fractures. Normal to the fractures, interference of scattered waves from the fractures results in a complex wavefield (Figure 6a). Fractures act as secondary sources with observable forward and backscattering as well as multiple scattering events. Parallel to the fractures the wavefield is much more coherent and the fractures appear to act as waveguides (Figure 6b).

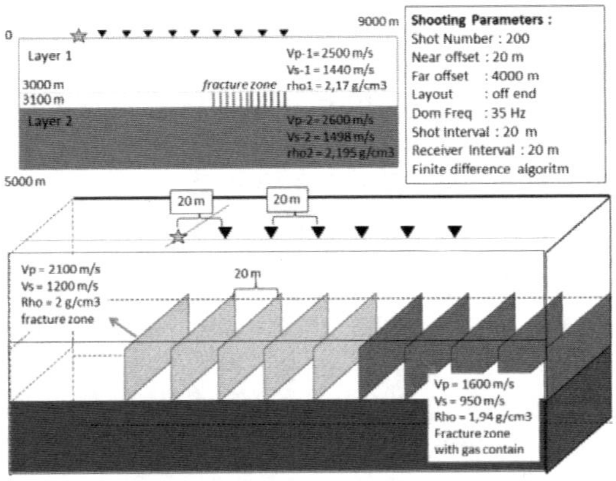

Figure 5. The model geometry for fracture zone showing source and receiver locations, fracture locations, and density parameters

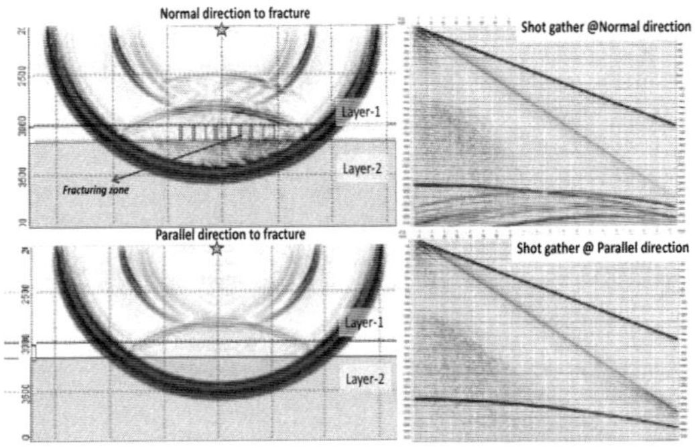

Figure 6. Snapshot of the vertical component of wave propagation in the normal (a) and parallel directions (b).

4. Result & Discussion

A CMP gather was generated using the elastic wave equation. The gather was NMO corrected using the initial stacking velocities estimated from the (vertical) interval velocities in the input model. The reflection event from the top of the fracture zone is flat and strong across all offsets. However, the reflection from the base of the fracture gas zone is only relatively flat up to mid-offsets. At the far offsets, the reflection event curves upwards indicating anisotropy in the fracture gas reservoir. VVO analysis was done on this simple CMP model over the fracture gas reservoir, as seen on Figure 7. Analysis of the hyperbolic moveout velocity was carried out on the base reservoir event using the LEC method and Equation 1. Computed offset-dependent RMS velocity increases with offset (Figure 7c).

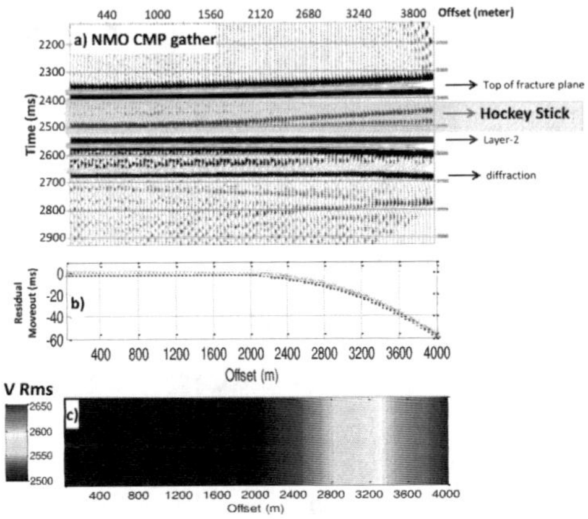

Figure 7. VVO attribute analysis at zone of interest.

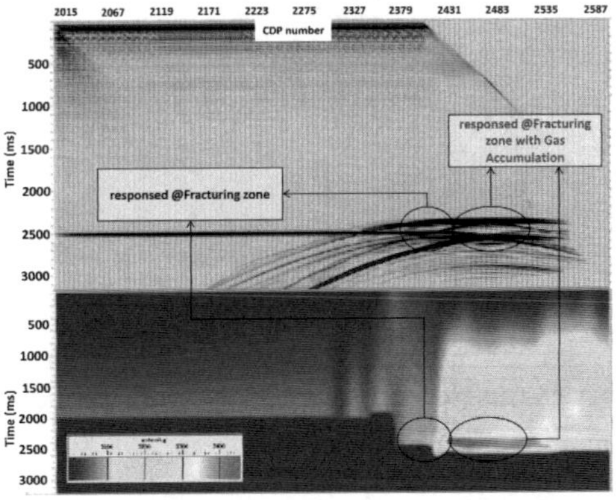

Figure 8. Stack section using Vrms derived from LEC and Vint gradient in colour

At location where the amplitude is bright, the *Vh* gradient is also positive (bright) as indicated by red to yellow colours. The *Vh* gradient is one product from direct velocity scanning of RMS velocity gathers. Before deriving the gathers for interval velocity *Vint*, the *Vh* gathers have to be converted to *Vint* gathers using a smoothing gradient equation. Linear regression is done on the *Vint* gathers to derive interval velocity gradient. Figure 8 shows the *Vint* gradient attribute across the seismic line. The *Vint* gradient also exhibits high positive anomalies localized at location of fracture reservoir zone (Higher interval velocity gradient).

IOP Conf. Series: Earth and Environmental Science **29** (2016) 012020 doi:10.1088/1755-1315/29/1/012020

5. Conclusion

The Vrms field produced by the LEC method is a high-resolution velocity field. This high-resolution velocity field is very useful, for instance for developing a velocity model for depth conversion and for pore pressure prediction. The VVO method is not affected by processing algorithms and parameterization, unlike the AVO method. Observations from VVO attributes of hyperbolic moveout velocity gradient and interval velocity gradient that fracture reservoir zone with gas accumulations is more easier to identified. Meanwhile, the existence and orientation fractured zone are recognized in FMI logs of study wells due to the overall good quality of electrical images which can be used as reservoir characterization and fracture determination as good as the results of prospect development.

References

[1] Bishop M G 2001 South Sumatra Basin Province, Indonesia: The Lahat/Talang Akar-Cenozoic Total Petroleum System

[2] De Coster G L 1974 The geology of the Central and South Sumatra Basins *P*roceedings Indonesian Petroleum Association 3rd Annual Convention p. 77-110

[3] Gray D, Roberts G, and Head K 2002 Recent advances in determination of fracture strike and crack density from P-wave seismic data *The Leading Edge* v. **22** p. 280-285

[4] Gudmundsson Agust 2011 *Rock Fractures in Geological Processes* Cambridge University Press, Cambridge, England, ISBN 978-0-521-86392-6

[5] Hutchison C S 1996 *South-East Asian Oil, Gas, Coal and Mineral Deposits*: Clarendon Press Oxford

[6] Minsley B J, Burns D R, and Willis M E 2003 *Fractured reservoir characterization using azimuthal AVO* ERL Industrial Consortium Annual Report MIT

[7] Nihei K T et al 2002 Finite difference modeling of seismic wave interactions with discrete, finite length fractures 72nd Ann. Int. Mtg. *Soc. Expl. Geophys* Expanded Abstracts

[8] Perez M A, Grechka V, and Michelena J 1999 Fracture detection in a carbonate reservoir using a variety of seismic methods *Geophysics* v. **64** p. 1266-1276

[9] Ruger A 1998 Variation of P-wave reflectivity with offset and azimuth in anisotropic media *Geophysics* **54** 680-688

[10] Sapiie et al 2011 Carbonate Fractured Reservoir Characterization Using Analogue Outcrop Study Of The Rajamandala Carbonate Complex, West Java *Proceedings Indonesian Petroleum Association*

[11] Schoenberg M and Sayers C M 1995 Seismic anisotropy of fractured rock *Geophysics* v. **60** p. 204-211

[12] Shen F et al 2002 Case History: Azimuthal offset dependent attributes applied to fracture detection in a carbonate reservoir *Geophysics* v. **67** p.355-364

[13] Shen F and Toksöz M N 2000 Scattering characteristics in heterogeneously fractured reservoirs from waveform estimation *Geophys. J. Int.* v. **140** p. 251-266

[14] Supriyono 2011 *Velocity Variation with Angle or Offset (VVO) A New Tool for Direct Hydrocarbon Indicator: Case Studies in Low and High Impedance Clastic Reservoirs* PhD dissertation Department of Geophysical Engineering Bandung Institute of Technology Bandung

[15] Suseno P H, Zakaria, Mujahindin, Nizar, and Subroto E A 1992 Contribution of Lahat Formation as hydrocarbon source rock in South Palembang area, South Sumatera, Indonesia *Proceedings Indonesian Petroleum Association Twenty First Annual Convention* p.325-337

[16] Wijaya O 2015 *Prediction of Fracture Porosity from Well Log Data by Artificial Neural Network, Case Study: Carbonate Reservoir*, "Depok" Field

Anisotropy of Magnetic Susceptibility (AMS) analysis for sedimentation tracing of Selorejo reservoir

S Zulaikah[1], R Azzahro[1], E S Mu'alimah[1], Y Bungkang[2]
[1]Department of Physics, Universitas Negeri Malang, Indonesia
[2]Department of Physics, Universitas Cendrawasih, Indonesia

E-mail: siti.zulaikah.fmipa@um.ac.id

Abstract. We use the Anisotropy of Magnetic Susceptibility (AMS) method to describe the source and mechanism of sedimentation process at the Selorejo reservoir. Six cores of sediment were collected by conventional coring process and sliced into sub samples in order to trace the pattern of sedimentation in the horizontal section. Based on the average of magnetic susceptibility of each core we are able to distinguish the origin of sediment from the water input into the reservoir. Our result agrees with the SEM images of the magnetic mineral grains, extracted from the inputs. However, the data shows the size difference from those two inputs about one order of magnitude. The average layer-by-layer Eigen-values of the susceptibility in horizontal section are consistent, which also shows consistency of the sedimentation process along the vertical direction.

1. Introduction

The detailed study of reservoir sedimentation is still rare due to its mechanism complexity. In this case the rate is higher compared to the lake or sea sedimentation [1, 2, and 3]. The reservoir sedimentation also characterized by a complex sediment yield. These factors make this type of sediments valuable, especially for a detailed study of properties of magnetic minerals, which may give some clues to understand processes and mechanisms that occurs in the environment [4, 5, and 6]. The reservoir watershed is usually used as a water supply, irrigation, hydropower and tourism object. This makes it very important for the surrounding community, to understand the sedimentation transport process and rate to maintain the sustainability of the reservoir.

Tracing the transfer processes of the sediments from the water input through the sedimentation area up to the output is one of the important studies [7, 8, and 9]. This process can be tracked using the Anisotropy of Magnetic Susceptibility (AMS) technique, which provides useful information regarding the transport of magnetic minerals dissolved in the sediment. In the last sixty years, AMS has been used in many applications. Among the important applications are study on the sediment fabrics and depositional process on the unconsolidated tsunami deposits [10, 12], describing the deformed sedimentation rocks [13], reconstructing paleo-monsoon route [14], tracing the signature of precipitation [15], and tracing the paleo-environmental changes [16, 17]. The method provides a rapid and precise estimation of the degree and direction of preferential alignment of magnetic minerals assemblages [8]. In this paper, we describe a detailed process on a special type of sediment taken from the specific area of our research location.

2. Sample Preparation and Method

Selorejo reservoir lies in Ngantang Malang East Java Indonesia (110° 30' - 112° 55' E and 8° 15' S). The location is about 48 miles in the northwest city of Malang, surrounded at least four mountains, namely mount Kawi, mount Anjasmoro, mount Arjuno and mount Kelud. The reservoir has two main water inputs, the upstream of Konto and Kwayangan River. The catchment area is about 235 km^2, with the maximum reservoir capacity is about 62.30 Mm^3. The effective reservoir capacity is 50.10 Mm^3.

We collected six cores of sediment that is distributed spatially for AMS analysis. We perform a conventional coring process from a sampling boat. The diameter of the cores is 2 cm and ranged from 10 cm to 32.5 cm depth. We distribute the core sampling locations to represent the important sites, which are the water input from the two rivers, Konto River (TP1, TP2) and Kwayangan River (TK1, TK2), the center of the reservoir (TT) and the water output towards the Pinjal River (DKTP). The location and the used notation are shown in figure 1. From each location we sliced 2.2 cm sub-samples, which were packed into 10 ml cylindrical plastic holder. The magnetic susceptibility of the samples was measured using the Bartington magnetic susceptibility meter (MS2B).

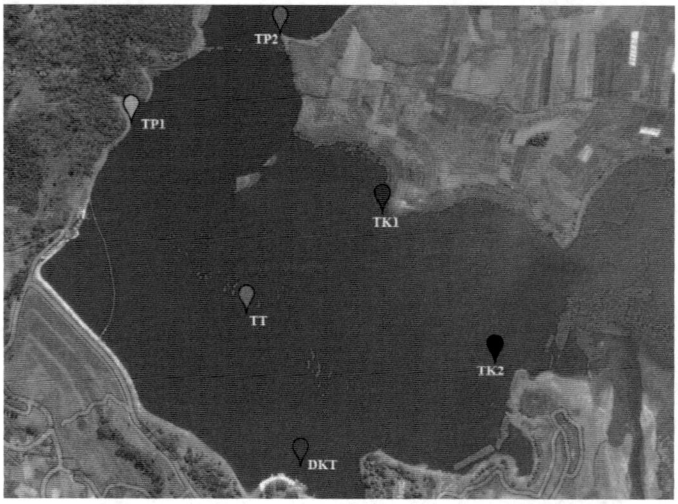

Figure 1. The sample core locations. The input rivers, the Konto and the Kwayangan, are marked by TP and TK, respectively. The output river, the Pinjal, is denoted by DKT. The sedimentation sites is marked by TT, at the center position of the reservoir. (Source: google map with some modification)

At low frequency, the susceptibility (χ_{lf}) was measured using AMS in 13 directions following the so-called Borradile method [18]. From this data the values of maximum-, minimum- and the intermediate-susceptibility can be determined. These values are the Eigen values of the magnetic susceptibility, regarding the AMS ellipsoid axes. We denote these values as K_{max}, K_{min}, and K_{int}, respectively. The analysis of every data is based on these AMS parameters, derived to additional parameters, namely, the average of magnetic susceptibility (K_{ave}), the lineation ($L=K_{max}/K_{int}$), the foliation ($F=K_{int}/K_{min}$), the degree of anisotropy ($P=K_{max}/K_{min}$), and the shape parameter ($T=[\ln F - \ln L]/[\ln F + \ln L]$). In rocks and sediments the anisotropic behavior commonly originates from the magneto-crystalline anisotropy, where the principal axes are related to the crystallographic axes of the mineral grain, and the shape anisotropy, where the elongated mineral grain is due to the self-demagnetizing factor in a specific direction [8].

3. Results and Discussion

The magnetic susceptibility of samples from two water inputs has a different property, where the value of TK is higher compared to that of TP, hence from the data, the origin of the magnetic minerals can be easily distinguished. This finding is also supported by the result of scanning electron microscopy and the dispersive spectroscopy (SEM/EDAX) analysis of the corresponding samples of the sediment reservoir. The EDAX analysis shows that the magnetic constituent is dominated by *titanomagnetite*, due to the existence of Fe-Ti-0 elements (c.f. table 1). This fact is also verified by the morphology from electron micrograph shown in figure 2. In case of *titanomagnetite* minerals the anisotropic behavior in AMS is caused by the shape effect, since this mineral is crystallographically isotropic [8]. The SEM images also demonstrates the variation of grain size of the samples taken from TK and TP sites, which determines two categories of sizes, which have one order of magnitude difference as shown in figure 3.

Tabel 1. Elements dissolved in representative sample of sediment. Fe is a major element.

Element	*Wt%*	*At%*
OK	28.74	52.72
MgK	02.13	02.57
AlK	09.10	09.89
SiK	05.61	05.86
TiK	04.13	02.53
FeK	50.30	26.43
Matrix	*Correction*	*ZAF*

Figure 2. The electron micrograph of *Titanomagnetite* mineral with defect in one side, shown by the red rectangle. The sub-sample was extracted from TT7 location.

Figure 3 shows the variation of AMS parameters along the vertical section of the core taken from TP1, represents the initial sediment samples, and core TT, that represents the sediment yield. We found no specific pattern of AMS parameter for both cores, in the vertical direction. However, we found the variation of the shape factor, which shows a wide range at the bottom and narrow range at the middle and the top. The value of foliation and lineation oscillates along the samples, which indicates the oblate and prolate grains, induced by the transport process. The maximum lineation indicates that the prolate grain and the sedimentation is dominantly affected by the transport process. Otherwise, the minimum value indicates the oblate grain, created by the compaction process.

(a) (b) (c)

Figure 3. The SEM imaging of magnetic minerals extracted from sub samples (a) TK2.7, (b) TP1 and (c) TT7, using 200x magnification. The figure sequence shows clearly the grain size gradation of magnetic mineral from the sample points.

Assuming that the dissolved minerals were transported from the input river to the output, we mapped the horizontal section to study the layer-by-layer sedimentation. For this purpose we fit the three AMS parameters; K_{ave}, the lineation and the foliation. We found two horizontal patterns, shown in figure 4. The plots show a consistent pattern along vertical section (shown in figure 5), which reveal the change in flow dynamics. This finding can also interpreted as the size difference of the grain originated from the Kwayangan River and the Konto River. The first transfers larger grains that mix in the center location of the reservoir. These grains become smaller at the output (Pinjal River).

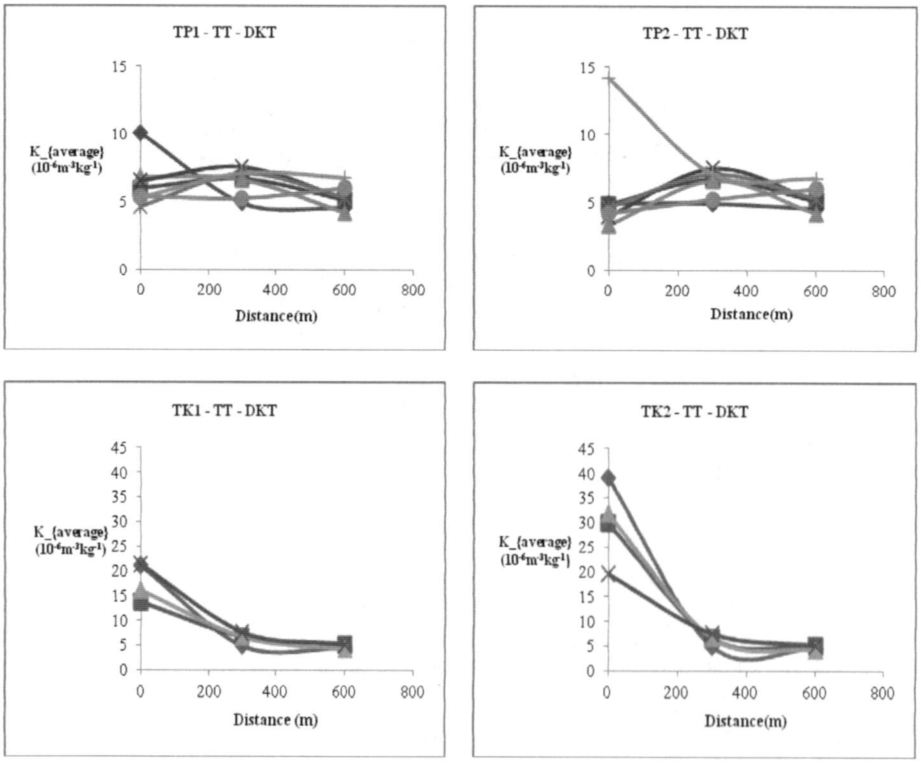

Figure 4. The variation of the average Eigen value, K_{ave}, as a function of a distance of magnetic mineral transport from the source in horizontal section (from TKI or TK2 to TT and ends at DKT, c.f. Fig. 1.)

IOP Conf. Series: Earth and Environmental Science **29** (2016) 012021 doi:10.1088/1755-1315/29/1/012021

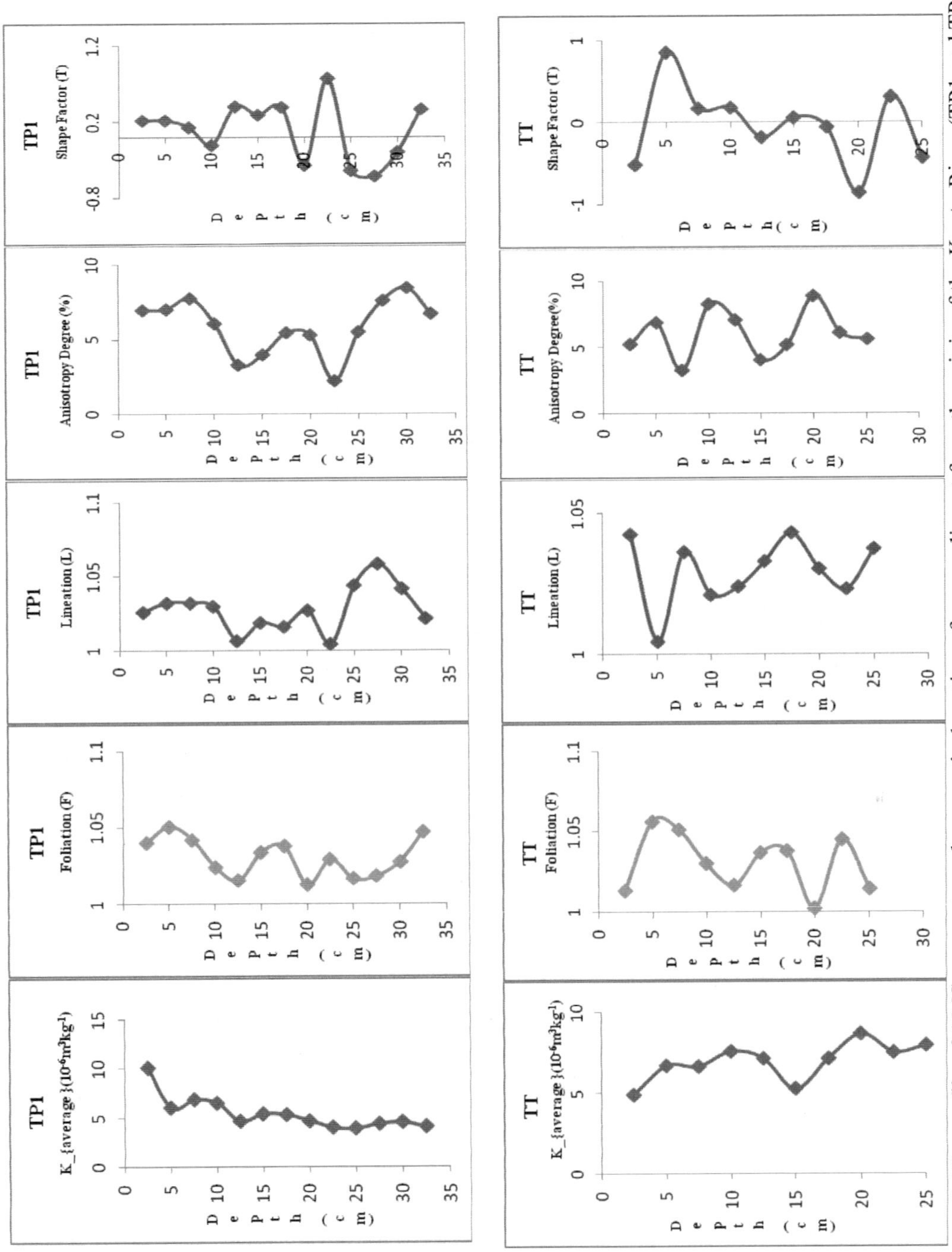

Figure 5. The profile of AMS parameters along vertical section of core sediment from the origin of the Konto River (TP1 and TP2) and the sediment at the center of reservoir (TT)

5. Conclusion

Based on the average of magnetic susceptibility of each core, we can distinguish the origin of sediment input. This result is in agreement with SEM images that show the grain size of magnetic mineral from the two cores taken from the Kwayangan and the Konto River as an input of water and sediment. The difference is about one order of magnitude. The average of the layer-by-layer magnetic susceptibility in horizontal section is consistent, represents consistency of the surrounding process along vertical section. Our result shows that foliation and lineation was developed alternately along the core, however this does not demonstrate a specific pattern.

Acknowledgements

The authors would like to thanks to Husni Cahyadi Kurniawan, Agus Riyanto, Shelita Sindaratna, Rizka Amirul Hikma, Lutfia Tri Wahyuni, Yuni Chairun Nisa and Susanti Mayangsari for grate help on the sampling process. Thanks also to PEDISGI2015 committee for immeasurable help of publishing this paper.

References:

[1] Loizeau J-L, Roze S, Peytremann C, Monna F, Dominik J 2003 Mapping Sediment Accumulation Rate by Using Volume Magnetic Susceptibility Correlation in contaminated Bay (Lake Geneva, Switzerlamd) *Eclogae geol. Helv* **96** Supplement 1, S73-S79.

[2] Tamuntuan G, Bijaksana S, Gaffar E, Russel J, Safiuddin L O, and Huliselan E 2010 The Magnetic Properties of Indonesian Lake Sediment: A Case Stdy of a Tectonic Lake in South Sulawesi and Maar Lakes in East Java *ITB J. Sci.* **V.42** No. 1 31- 48.

[3] Baltrunas, V., Seirien, V., Molodkov, A., Zinkut, R., Katinas, V., Karmaza, B., Kisielien, D., Petrosius, R., Taraskevicius, R., Piliciauskas, G., Schmölcke, U., and Heinrich, D., 2013, Depositional environment and climate changes during the late Pleistocene as recorded by the Netiesos section in southern Lithuania, Quarternary International, V.**292**, 136-149.

[4] Campos C, Beck C, Crouzet C, Carrillo E, Welden AV, and Tripsanas E 2013 Late Quaternary paleoseismic sedimentary archive from deep central Gulf of Corinth: time distribution of inferred earthquake-induced layers *Annals of Geophysics* **56** 6

[5] Murdock K J, Wilkie K, and Brown L L 2013 Rock magnetic properties, magnetic susceptibility, and organic geochemistry comparison in core LZ1029-7 Lake El'gygytgyn Russia Far *East Climate of The Past* **9** 467-479.

[6] Pozza M R, Boyce J I, and Morris W A 2004 Lake-based magnetic mapping of contaminated sediment distribution, Hamilton Harbour, Lake Otario, Canada *Journal of Applied Geophysics* **57** 23-41

[7] Hatfield H G, and Maher B A 2009 Fingerprinting Upland Sediment Sources: particle size-specific magnetic linkages between soils, lake sediments and suspended sediments, *Earth Surface Process and Lanforms*

[8] Hailwood E A and Sayre W O 1983 Sediment Transport Mechanisms at NE Atlantic Margins: Evidence from the Magnetic Anisotropy of Ipod Cores, Structure and Development of the Greenland-Scotland Ridge Nato Conference Series **Volume 8** pp 479-493

[9] Bungkang Y and Soemarno 2014, Sentani Watershed Erosion Potential Study and Suspeded Solid Distribution In Connection With Lake Silting, *Journal of Environmental Science, Toxicology and Food Technology* (IOSR-JESTFT), **V.8**

[10] Paris R, Wassmer P, Lavigne F, Belousov A, Belousova M, Iskandarsyah Y, Benbakkar M, Ontowirjo B, and Mazzoni N 2014 Coupling eruption and tsunami records: the Krakatau 1883 case study, Indonesia, Bull Volcano **76** 814

[11] Horton B, MacInnes B, Gonzalez F, Hemphill-- - Haley E, Switzer A, Witter R, Tanioka Y, and Pilarczyk J, ----, Long-Term Records of Tsunamis (and Storms) with Insights from Recent Events

[12] Veerasingam S, Venkatachalapathy R, Basavaiah N, Ramkumar T, Venkatramanan S, and Deenadayalan K 2014 Identification and characterization of tsunami deposits off southeast coast of India from the 2004 Indian Ocean tsunami: Rock magnetic and geochemical approach *J. Earth Syst. Sci.* **123** pp. 905–921

[13] Pares J M 2015 Sixty years anisotropy of magnetic Susceptibility in deformed sedimentary rock *Review Article Frontiers in Earth Science*

[14] Zhang R, Kravchinsky V A, Zhu R, Yue L 2010 Paleomonsoon route reconstruction along a W–E transect in the Chinese Loess Plateau using the anisotropy of magnetic susceptibility: Summer monsoon model *Earth and Planetary Science Letters* **299** 236-246.

15] Pronovost A L, St-Onge G, Gogorza C, Jouve G, Francus P, and Zolitschka B 2014 Rock-magnetic signature of precipitation and extreme runoff events in south-eastern Patagonia since 51, 200 cal BP from the sediments of Laguna Potrok Aike *Quaternary Science Reviews* 98 110-125

[16] Sangren P and Snowball I 2001 Application of Mineral Magnetic Techniques to Paleolimnology in Tracking Environmental Change Using Lake Sediments Physical and Chamical Technique **V.2** Kluwer Academic Publisher Dordrocht, The Netherlands

[17] Shouyun H U Chenglong D Appel E and Verosub K L 2002 Environmental Magnetic Studies of Lacustrine Sediment *Chinese Science Bulletin* **V.47** No 7

[18] Borradaile G J 1995 Anisotropy of Magnetic Susceptibility: Measurement Schemes *Geophysical Research Letter* **22**

IOP Conf. Series: Earth and Environmental Science **29** (2016) 012022 doi:10.1088/1755-1315/29/1/012022

Paleoecology and Paleoclimate of Tanjung Formation Deposition, Based on Palynological Data From Siung Malopot, Central Borneo

Kristyarin D A [1], A T Rahardjo , Bambang P
[1] Geological Engineering Department, Faculty of Earth Sciences and Technology, Institut of Technology Bandung, Jalan Ganesha 10, Bandung, Indonesia, 40132

E-mail : kristyarin@gmail.com,

Abstract. The northern part of Barito basin is exposed around Siung Malopot area (Central Kalimantan), showing the late cretaceous basement of andesite lavas and granite intrusion. The basement is unconformably overlied by the rocks of Tanjung formation at middle to late Eocene age. Compressive phase during Miocene and Plio-Pleistocene folded and uplifted all existing rocks, made the rocks become relatively north-south structurally oriented. The exposed Tanjung formation consists of sandstone, mudstone, siltstone with intercalation of coals and thin layers of limestone. These lithologies may represent the upper parts of Tanjung formation. Palynology quantitative study is an approach to identify the palaeoclimate and palaeoecology of a certain sedimentation system. Analyses from 18 samples taken from lithologies of Tanjung formation show the presence of palinomorph fossils associated in *Proxapertites operculatus* zonation indicating the P18-P20 age intervals, or late eocene age. The ratio of spore and non-spore percentage, as well as the comparison between arboreal pollen (AP) and non-arboreal pollen (NAP) show the significant change of environment from time to time during the deposition of the Tanjung formation, also indicating the climate humidity that became more stabile to the younger ages. From 52 observed taxa in the samples, the ratio of marine versus non-marine taxon indicates the influence of sea-water influx in the sedimentation system, explaining the possible presence of thin limestone intercalations at the upper parts of Tanjung formation.

Palynology quantitative study on Tanjung Formation indicates the sedimentation of this formation was in backmangrove environment that tends to become more influenced by marine conditions.

1. Introduction

Borneo island has a complex geological history, as can be clearly seen from the diversity of basement lithology as a result of tectonic events at Pre-Tertiary age. The island is composed of varied amalgamation basement that preserved in cratonic area, then metamorphed and folded during the pre-tertiary uplift [1]. Potential energy resources, such as coal, are also found abundantly in the Borneo island, as in the Tanjung formation in the Barito basin. Tanjung formation along with basement rock are exposed in the research area which is in the northern part of Barito basin, Siung Malopot area, North Barito regency, Central Kalimantan. The coordinates of research area is mN 9821000 - 9828500 mE mN and 302 500 - 308 500 mE (UTM zone 50 S), within 277 km to the northeast from Palangkaraya. Surface geological mapping study was conducted to renew the regional geological map that was done by [2].

Stratigraphy of research area was divided into four unofficial lithology unit, from older to younger are andesit unit (equivalent to Pitap formation), granodiorite unit (equivalent to cretaceous granite group), sandstone-mudstone unit (equivalent to Tanjung formation), and alluvial deposit unit (Figure 1). Geological structure in research area such as fold, thrust fault, and shear fault were developed after the deposition of sandstone-mudstone unit at eocene age. The compression regime is expected to begin at early miocene and reached the peak at plio-pleistocene, along with the tectonic activity in the Barito basin which was associated with Meratus uplift.

Based on the previous research interpretation, Tanjung formation was deposited in the transitional environment with many land plants as the main source of coal formation, which these plants can leave as residual organic material in sedimentary rocks. According to this interpretation, the authors were interested in conducting a palaeoecology and paleoclimate study based on palynological data. Palynology is the study of palynomorph fossils, including pollen, spores, dinoflagellate, acritarch, chitinozoa, and scolecodont, as well as organic materials and kerogen that is found in the sedimentary rocks and sediments [3]. The purpose of this analysis was to determine the depositional environment condition and determine the environmental change during the deposition of Tanjung formation that exposed in the research area.

2. Data Acquisition and Analysis Method

Rock samples required for palynological analysis were fine grained sedimentary rocks (such as clay, silt, shale, silty sand) which has dark color as an obvious indication of organic material in sedimentary rocks. These rock samples were taken at certain points while conducting stratigraphy measuring section (Figure 2), as many as 18 sample from top to bottom of the sedimentary rock outcrop layers. Stratigraphy measuring section was done by measuring the actual thickness of Tanjung formation outcrops that well exposed at the edge of Hasnur road. The samples were then prepared in the Palynological Laboratory in Institute of Technology Bandung to separate the palynomorph fossils with other elements of sedimentary rocks.

The method used in this research was quantitative analysis method, based on calculation of every observed palynomorph, with the following stages:

2.1 Determination Stage

Palynomorph determination was done by palynomorph observation through binocular microscope with 400 times and 1000 times magnification. The goal of this stage was classifing every palynomorph up to the level of family, genus, or species based on its physical properties (such as aperture, size, shape, and ornamentation). Each individual of determined palynomorph was counted and recorded accurately in number.

2.2 Palynomorph Classification Stage

At this stage, the determined palynomorph data taken during the determination stage was classified based on its environment vegetation according to [4]

IOP Conf. Series: Earth and Environmental Science **29** (2016) 012022 doi:10.1088/1755-1315/29/1/012022

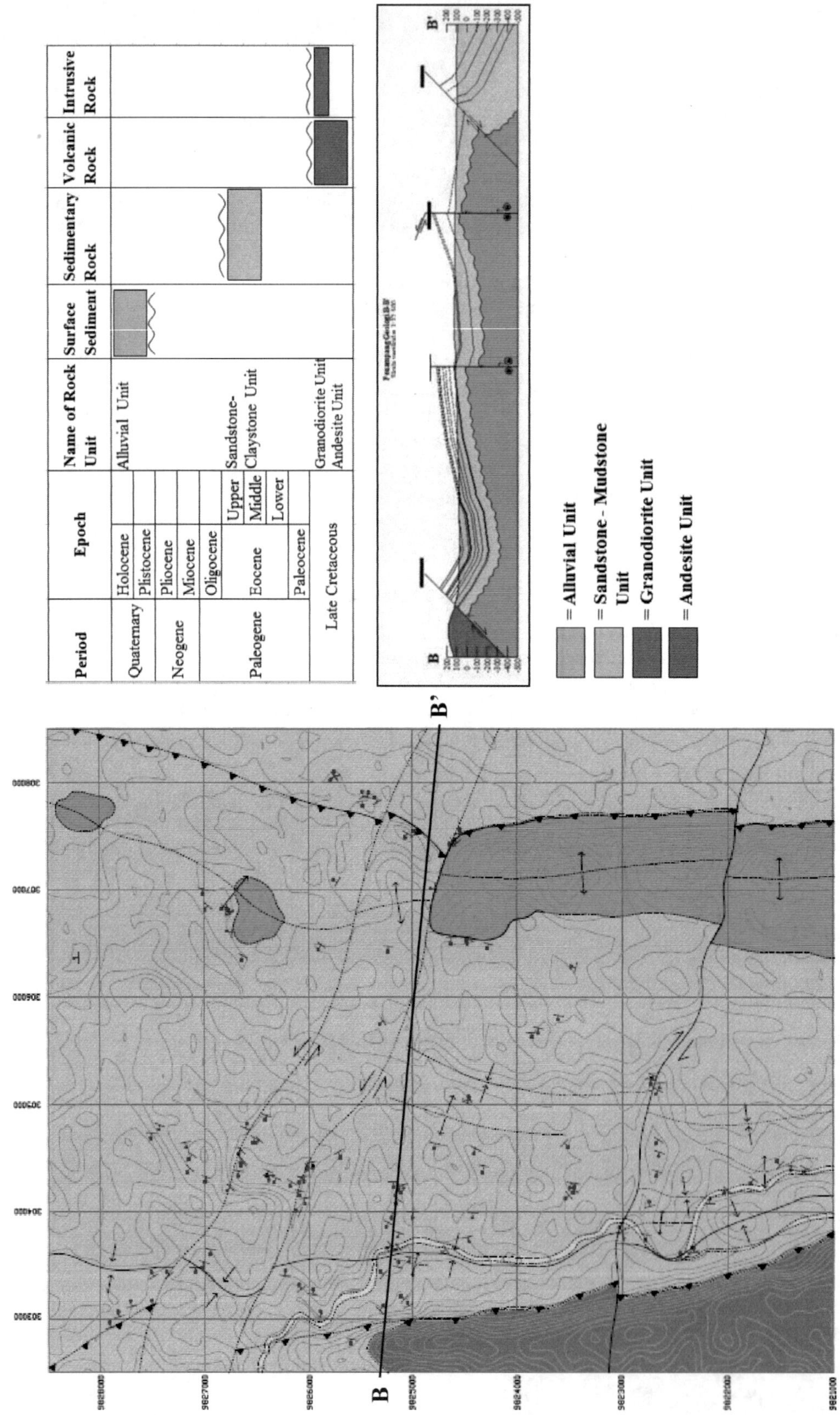

Figure 1. Geological map of research area

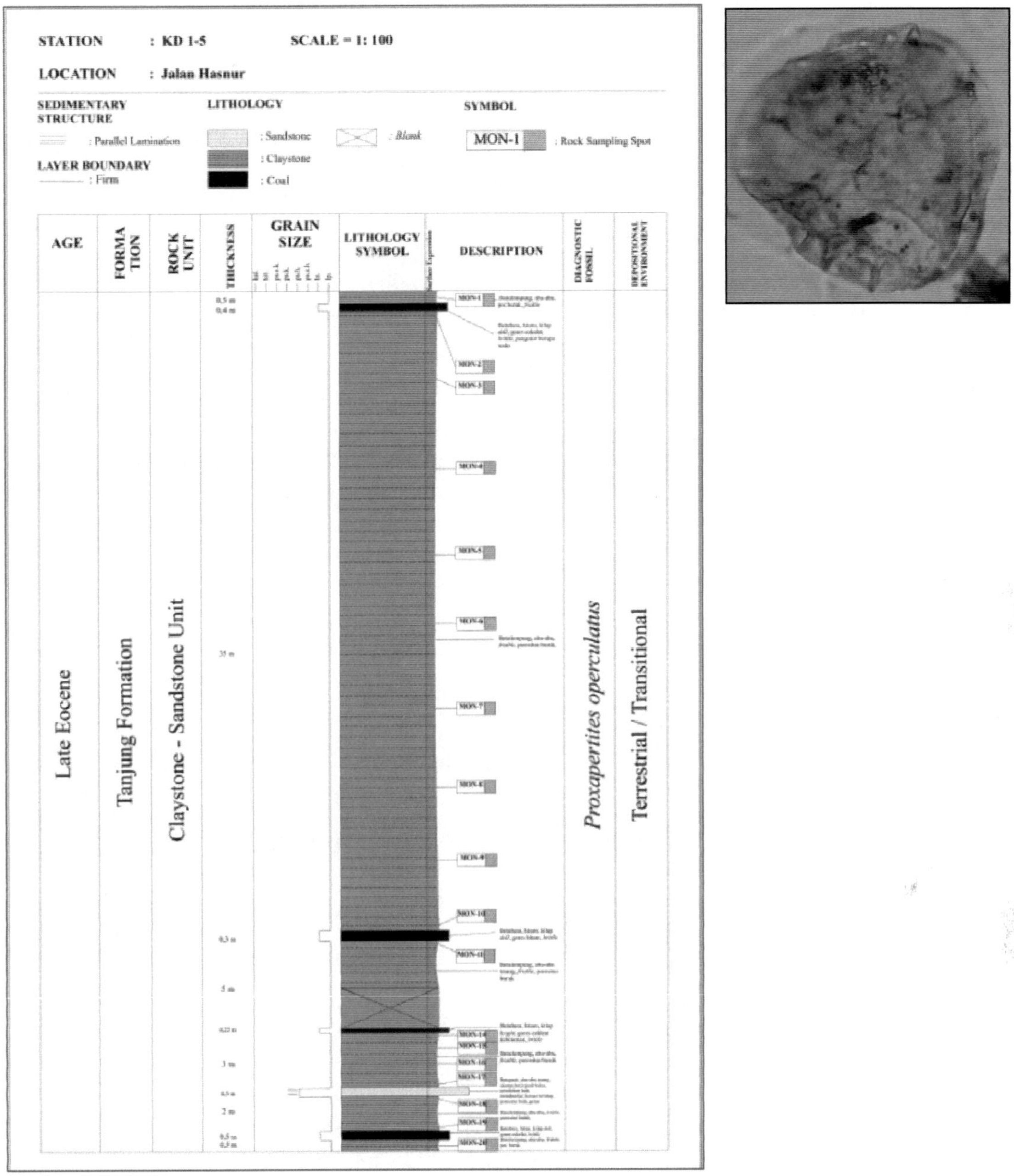

Figure 2. Stratigraphy measuring section column, based on the field observation at the outcrop on Jalan Hasnur (left) and the photo of *Proxapertites operculatus*, index fossil of eocene age (right)

2.3 Data Presentation
The palynological data is presented as palynological diagram. The calculation formulas for palynological diagram are as below:

2.3.1 Palynomorph Proportion Percentage
The percentage of each type of palynomorph can be obtained by comparing the amount of a group of palynomorph with the total number of palynomorphs in each sample (pollen, spores, fungi,

dinoflagellate and foraminifera tests lining). The proportion of a palynomorph group in this diagram illustrates the proportion of plants found in research area during the deposition of sandstone – mudstone unit. Here is the formula of a palynomorph proportion percentage :

$$\% \text{ a taxa} = \frac{\sum \text{a taxa}}{\sum \text{all palynomorph}} \text{ x } 100\ \% \tag{1}$$

2.3.2 Spore / Pollen Percentage

This percentage would give an idea about palaeoclimate according to the humidity level at the past time. The number of spores percentage represents the rate of humidity in palaeoclimate, which leads to interpretation whether the palaeocology was in dry or wet climate at the certain time. The percentage of spores is a result of comparison of a number of Pteridophyta spores by the total number of all palynomorph (excluding foraminifera tests lining and dinoflagellate), where the formula can be presented as follows:

$$\% \text{ spore} = \frac{\sum \text{spore}}{\sum(\text{spore+pollen})} \text{ x } 100\ \% \tag{2}$$

2.3.3 *Arboreal Pollen (AP) / Non-Arboreal Pollen (NAP) Percentage*

Arboreal pollen is pollen which was derived from woody plants as part of a forest that is identical with closed environment. Non-Arboreal pollen is pollen of graminae, herbaceae, and plants that live in an open environment (such as savannas and steppes). The percentage of AP / NAP illustrates the forest environment boundary movement as a reflection of environmental change. Here is the formula to determine the percentage of AP / NAP :

$$\% \text{ NAP} = \frac{\sum \text{NAP}}{\sum(\text{AP+NAP})} \text{ x } 100\ \% \tag{3}$$

$$\% \text{ AP} = \frac{\sum \text{AP}}{\sum(\text{AP+NAP})} \text{ x } 100\ \% \tag{4}$$

2.3.4 Marine / Non Marine Taxa Percentage

This percentage can be used to determine the movement of the shoreline to estimate the influence of marine environment during sedimentation, which can be formulated as follows:

$$\% \text{ Marine Taxa} = \frac{\sum \text{Marine Taxa}}{\sum(\text{Marine Taxa +Non-Marine Taxa})} \text{ x } 100\ \% \tag{5}$$

$$\% \text{ Non-Marine Taxa} = \frac{\sum \text{Non-Marine Taxa}}{\sum(\text{Marine Taxa+Non Marine Taxa})} \text{ x } 100\ \% \tag{6}$$

3. Age and Depositional Environment Analysis of Tanjung Formation

Based on the analysis of palynological diagram (Figure 3), the age of deposition sandstone-mudstone unit is late eocene. It is according to the presence of index fossils like *Proxapertites Operculatus*, and the presence of other fossils such as *Proxapertites cursus*, Proxapertites sp., *Meyeripollis naharkotensis*, *Cicatricosisporites eocenicus*, *Cicatricosisporites dorogensis*, and *Palmaepollenites kutchensis* that indicate that the sandstone-mudstone unit are in Proxapertites operculatus zone which isin P18-P20 age or late eocene [5]

In general, vegetation environment of sandstone-mudstone unit was backmangrove, proven by the abundance of *Acrosticum auerum* with average number in each sample reaching 48.59% (Figure 3). According to [4], this vegetation environment associated with depositional environment at the lower delta plain (Figure 4)

Palaeoecology and palaeoclimate condition at the deposition of sandstone-mudstone unit time was generally interpreted as the closed environment with high humid climate, evidenced by the average assemblage of AP (Arboreal Pollen) is 71.66% and the average assemblage of spores reached 68.82% in each sample. Nevertheless, there was the influence of the sea water flux, characterized by the presence of little amount of marine taxa percentage, with the average assemblage up to 2.46% per sample.

IOP Conf. Series: Earth and Environmental Science **29** (2016) 012022 doi:10.1088/1755-1315/29/1/012022

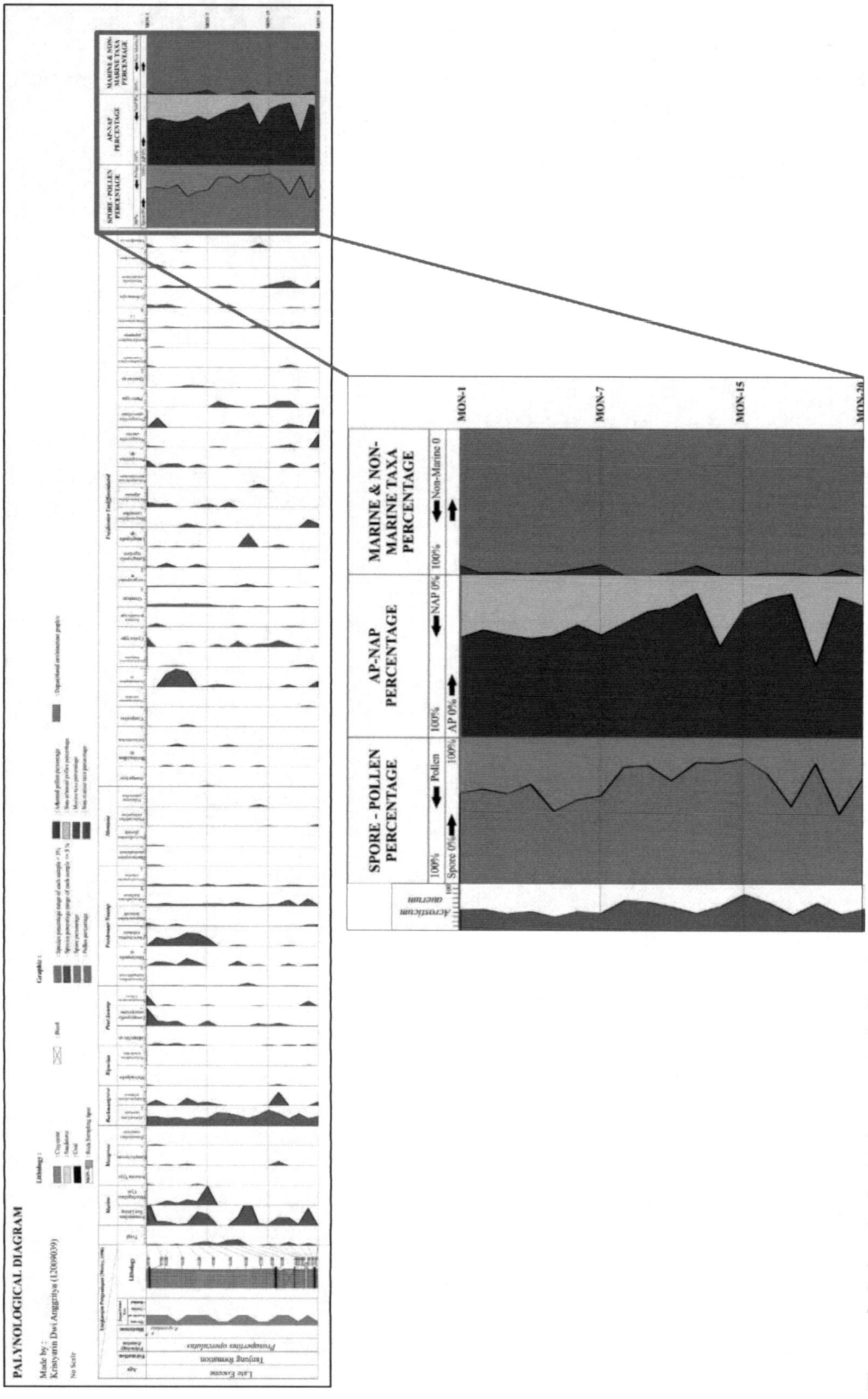

Figure 3. Palynological Diagram, including Spore-Pollen percentage chart, AP/NAP percentage chart and Marine / Non Marine Taxa percentage chart

4. Palaeoecology and Palaeoclimate Analysis

To simplify the palaeoecology and palaeoclimate analysis explanation based on the spore-pollen percentage chart, AP/NAP percentage chart and Marine / Non Marine Taxa percentage chart in palynological diagram (Figure 3), the graph was divided into three time deposition intervals from the older to the younger.

At the interval A, based on the AP/NAP chart and Spore/Polen chart, the fluctuation of palaeoecological changes during that period can be obviously seen. However, the opposite trend is shows by the palaeoecological and palaeoclimate conjunction : the closed environment associated with the low humidity environment (dry climates) and vice versa. The number of palinomorphs *Acrosticum aureum* was found very abundant at this interval, reached the highest number over all intervals at 79.09%. This kind of palinomorph indicates the vegetation environment of backmangrove [4]. There was a little effect of the sea water influx at the interval A, evidenced by the average percentage of Marine-Non Marine only at 1.5%.

Unlike the interval A, the graphic trend between the AP/NAP chart and Spore/Polen chart shows a relatively trend, they both show the slightly decreased trend from the older to the younger age at the same percentage level. The result of this comparison indicates the gradual environmental change from the closed environment into more open environment and also the gradual change from the dry climate into wetter climate at the same time. *Acrosticum aureum* is still found abundantly at the interval B with an average percentage of 51.82%, showing the same vegetation environment as interval A, backmangrove. There was an increase of marine taxa palinomorph number at this interval, where the average percentage of marine taxa increased up to 2.92% and foraminifera test lining taxa reached the peak of its development in this interval (7.02%). It is estimated that the marine influence towards to depositional environments is greater than at interval A.

The AP/NAP chart and Spore/Polen chart at interval C remained relatively constant, shows stable palaeoecological and palaeoclimatical change during that interval. There was a tendency of environmental changes into more open environment and more dry climate, compared to the two previous intervals. Depositional environment at this interval was still in backmangrove environment based on the highly abundance of *Acrosticum aureum*, but the number was steadily decreasing. The average percentage of marine taxa in this interval reached 2.94%, slightly increased compared to the previous interval. Interval C obtained the highest marine influence compared to the two previous intervals.

Based on the percentage of *Acrostichum aureum* and the percentage of marine taxa, there was a tendency of environmental changes from terrestrial environment to transitional environment from the older to the younger strata, but this condition could change fluctuatively at the certain time, as it have happened to the interval A. This tendency can explain the insertion of thin limestone in sandstone-mudstone unit. We interpret this environmental changes as a result of the sea level rising.

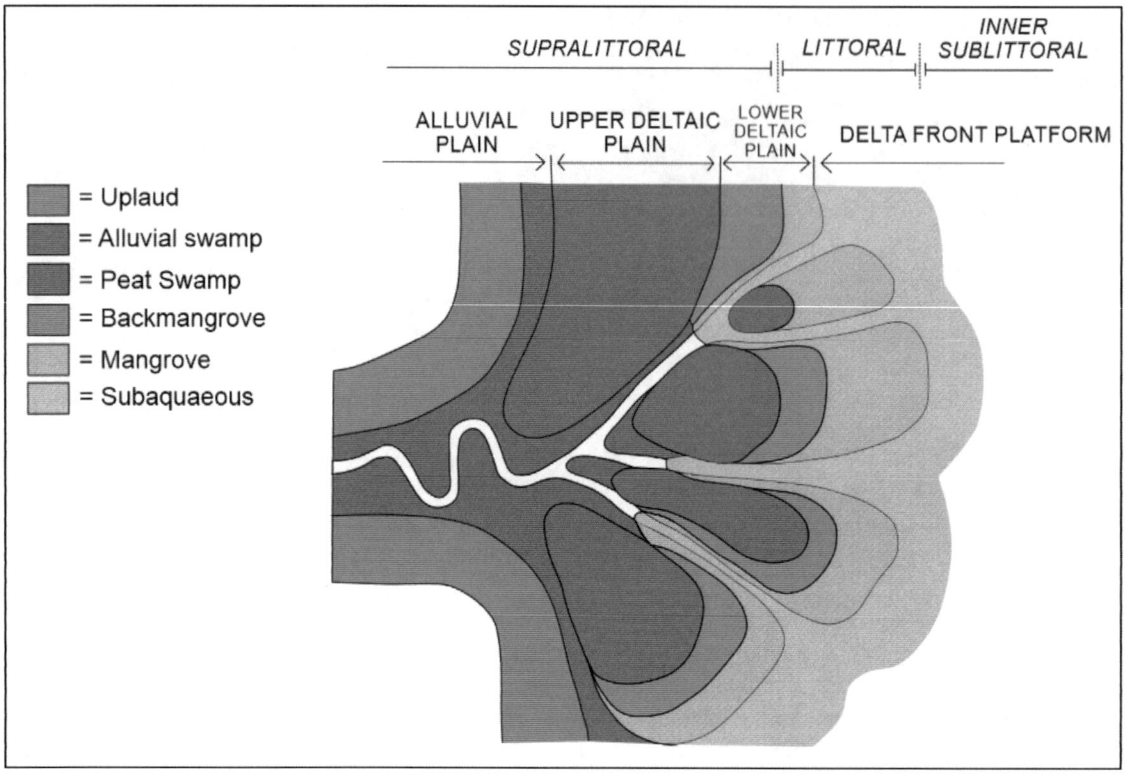

Figure 4. Relationship between vegetation environtment and depositional environtment, based on plant habitat (Modified from [4])

5. Conclusion

Sandstone-mudstone unit synchronized by the authors as Tanjung formation that was deposited at late eocene age (P18-P20). Vegetation environment of this unit in general was backmangrove that associated with lower delta plain depositional environment. This analysis also shows the palaeoecological condition of research areas was largely covered with forest as a closed environment with high humidity (wet climate). The environmental changes from the older to the younger deposition of sandstone-mudstone unit was relatively stable, although fluctuation is happened at certain interval, but it also shows a tendency to be more open and dry environtment.

The percentage of *Acrosticum aureum* was inversely proportional to the percentage of marine taxa, which indicates a change of the terrestrial environment into the transitional environment, from older to younger interval. This may explain the insertion of thin limestone that are present in upper Tanjung formation in research area.

6. Acknowledgement

This paper was presented at Padjadjaran Earth Dialogue: International Symposium On Geophysical Issues (PEDISGI), Padjadjaran University, 2015. The author would like to thank the committee of PEDISGI for allowing this paper to be published on the IOP Conference Series: Earth and Environmental Science. We also want to thank Indika Energy Tbk, which had funded this research. In addition, thanks aimed to the entire lecturer in Geological engineering Department ITB for their help and support during the research.

References

[1] Hall R, van Hattum M W A and Spakman W 2008 Impact of India – Asia Collision on SE Asia : The Record in Borneo, *Tectonophysics* **451,** 366 – 389
[2] Soetrisno, Supriatna S, Rustandi E, Sanyoto P and Hasan K 1994 Geological Map of Buntok Area, Borneo, Scale 1 : 250.000 (Indonesia : Geological Research and Development Center (PPPG))
[3] Traverse A 1988 Paleopalynology (London : Department of Geoscience, College of Earth and Mineral Science, The Pennsylvania State University), 5-15
[4] Morley R J 1990 Short Course Introduction to Palynology With Emphasis on Southeast Asia (Indonesia : Faculty of Biology of UNSOED University), 9-29.
[5] Rahardjo A T, Poulhaupessy A A, Wiyono S, Nugrahaningsih L and Lelono E B 1994 Pollen Zonation for The Tertiary of Java Island, Proc. 23rd Ann. Conv. Indon. Assoc. Geol. (IAGI), 77-87

Seismic reflection survey at Ayer Hangat site to investigate shallow subsurface structures

Amin E. Khalil[1, 2], **Mohd Nawawi**[1] **and Rami Kamel**[1]
1 School of Physics, Universiti Sains Malaysia, Pulau Penang, Malaysia, 11700
2 Geology Dept., Faculty of Science, Helwan Univ., Cairo, Egypt, 11795.

E-mail: amin_khalil@usm.my

Abstract. Ayer Hangat site is located in the island of Langkawi, northwest Malaysia. The site is characterized by the presence of hot spring. This hot spring is believed to be related to granitic intrusion nearby. Hence the present work is focusing on defining the shallow subsurface structures that control the migration of hot water to the surface. Seismic reflection method is used to achieve the goal of the present study. Forty three shot points were used with an offset of 5m of the nearest geophone. The shot-points interval is set to 1m. Seismograms were recorded on 24 channel TERRALOC instrument. The Geophone interval used was 1m. Conventional seismic data processing scheme was adopted. However, due to the fact that TERRALOC produce SEG2 data files, a script based on Obspy was written and used to convert to SEG-Y format. Afterwards, analyses were carried out using SU Package. The processed data is used to develop a model for the subsurface controlling structures. Such model will help in the understanding of the geothermal hot spring system in the area.

1. Introduction

Seismic reflection method is usually used for hydrocarbon exploration at relatively deeper depths. In that direction, the method gives high resolution images of the target structure relative to other geophysical techniques. For shallow depths, however, the method is less used and apparently losing ground for other geophysical techniques including seismic refraction method. The reason for that may be due to the extensive data processing labor required to produce meaningful seismic section. This doesn't tell all the story, some efforts were exerted to use seismic reflection to explore shallow depths adopting a simplified technique called optimum offset (e.g. [1], [2] and [3]). In the present work we will use that technique with some additional processing steps to enhance the resulting model.

The method is applied to Ayer Hangat complex site. The site is hot spring complex that currently undergoing development. The present study is applied there among other studies in order to help identifying the presence of hot water aquifer or pathway to the surface. The basic objective of the present work is to identify the near surface structures that may be responsible for water seepage to the surface. The structures are believed to be either faults or chimney like structures. Determination of such structure may help in decision on where to drill and where to build certain facilities.

Data processing in the present survey is carried out using both OBSBY [7] and Seismic Unix ([8] and [9]). Both are free and open source software. The use of OBSPY is restricted to preparing the field data by removing the DC-offset and to convert the original SEG-2 format into SEG-Y format. The rest of data processing are carried out using Seismic Unix packages on Ubuntu 14 'trusty' version.

2. Site and Geology

The area under investigation is located at Langkawi Island in the north western peninsular Malaysia (Figure 1). Geographically, the area of investigation lies at latitude 6° 25' 20.75" N and longitude 99° 48' 46.52" E. The surface is rather flat at the site with the presence of pavement rocks and channels as parts of the complex. This features posed some problems during data acquisition stage and some measures were followed to avoid their negative effects. Langkawi itself is a Geopark that comprises 99 islands that form what is known as legendary archipelago [4].

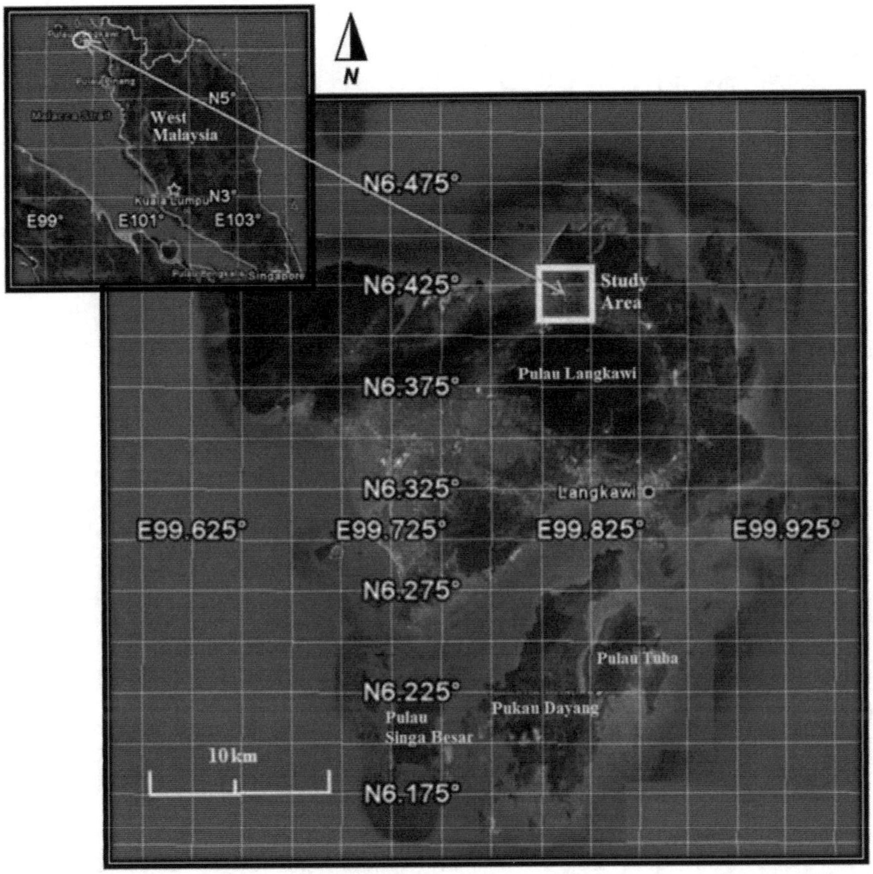

Figure 1. Location map of Ayer Hangat site.

The geology of Langkawi is a result of long depositional history under the various paleo-environmental conditions, followed by the tectonic and magmatic events. The principal geologic units (Figure 2) encountered in Langkawi Islands are listed below in reverse order ([5] and [6]):

6. Superficial Deposits.
5. Gunung Raya Granite (Late Triassic).
4. Chuping Formation (Middle Permian to Early Triassic).
3. Singa Formation (Late Devonian and Early Permian).
2. Setul Formation (Early Ordovician to Early Devonian).
1. Machinchang Formation (Late Cambrian).

The Machinchang Formation is considered to be the oldest sedimentary rock succession in Langkawi. This formation covers the northwest side of Langkawi. Machinchang Formation is a deltaic sequence, and it is composed of grey, brown, purple, and red, coarse to medium grained quartzite, arkose, and sub-greywacke [5].

Setul Formation is conformably overlying the Machinchang Formation, and exhibits a long period of marine transgression when the sediments deposited in relatively shallow marine environments ([6] and [4]). This formation composed of hard, brittle, dark-colored, thick-bedded limestone, with subordinate detrital facies [5]. The outcrops of this formation can be found on the eastern limb of the Machinchang anticline, and appear as a continuous belt along the eastern coast of Langkawi Island.

The Singa Formation rests unconformably on the Setul Formation, and it is surrounding the east and the south sides of the Gunung Raya granite dome in the middle of Langkawi Island [6]. Singa Formation deposited in the shallow marine environment, and it is composed of dark-grey or black well-bedded carbonaceous flagstones, grey and black shale, and siltstones with subordinate grey and brown immature quartzites and sub-greywackes [5].

Large and small igneous stocks are found scattered over Langkawi Island. The largest one found in the middle of Langkawi Islands called the Gunung Raya granite dome. The granite intruded into all older sedimentary rocks formations turning them into several types of metamorphic rocks.

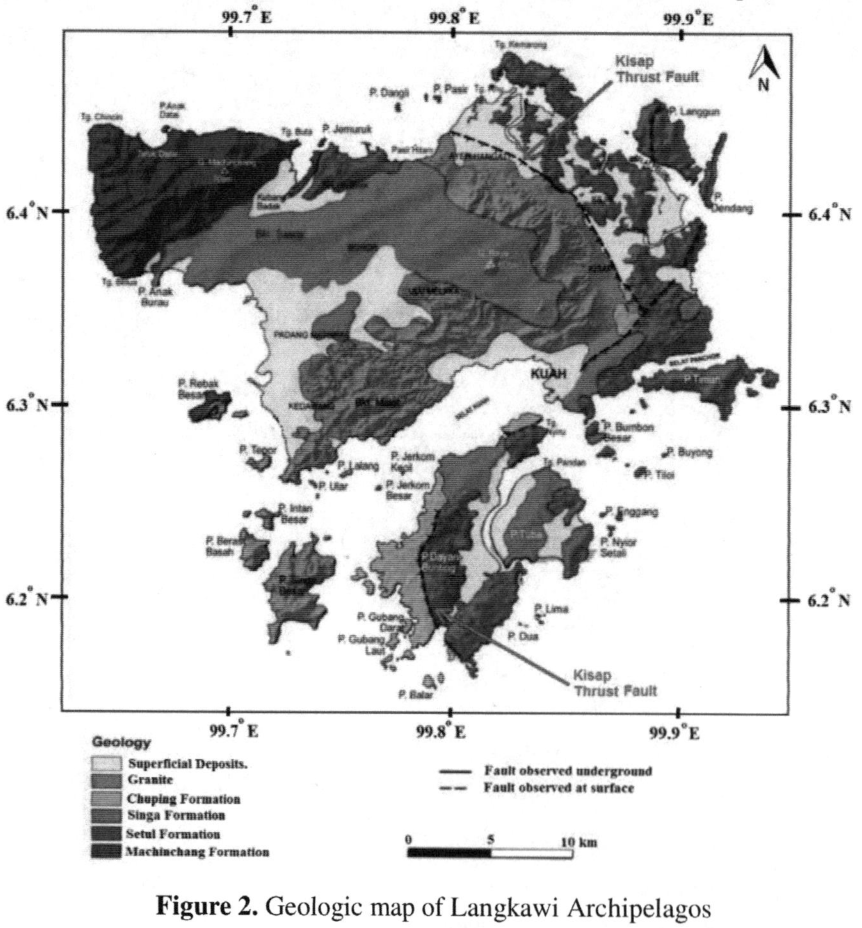

Figure 2. Geologic map of Langkawi Archipelagos

3. Methodology and data analysis

The method used in this survey can be termed as common offset method (e.g. [1] and [2]). In this method we tried to imitate the optimum offset method using data recorded at 24 geophones land streamer. Different common offsets were investigated in order to choose the one that produced the best resolution among the 24 geophones recorded for each shot. Subsequent processing techniques are then applied to enhance amplitude of the reflected data.

3.1 Data Acquisition

Seismic survey was carried out using land streamer with 24 geophones at 1 m interval. The source is 5 kg hammer at an offset of 3 m from the nearest geophone. This produce a maximum offset of 26 m. For the survey we used 48 shots points at 1 m offset. Hence the total length of the data acquired is 70 m (Figure 3). The first shot is located at the hot spring site. Seismic data are recorded by TERRALOC MK-8 instrument at sampling interval of 0.0001 sec. The total length of each recorded trace is 0.8192 seconds. Data files produced by TERRALOC are exported in special SEG-2 format. When trying to convert these file formats with *seg2segy* code accompanied with SU software we encountered problems and conversions could not be made. Another problem with the data recorded by TERRALOC is the DC-shift. Such shift must be removed before applying any data processing steps because it produced Gibb's phenomena due to the sharp change in amplitude at both start and end.

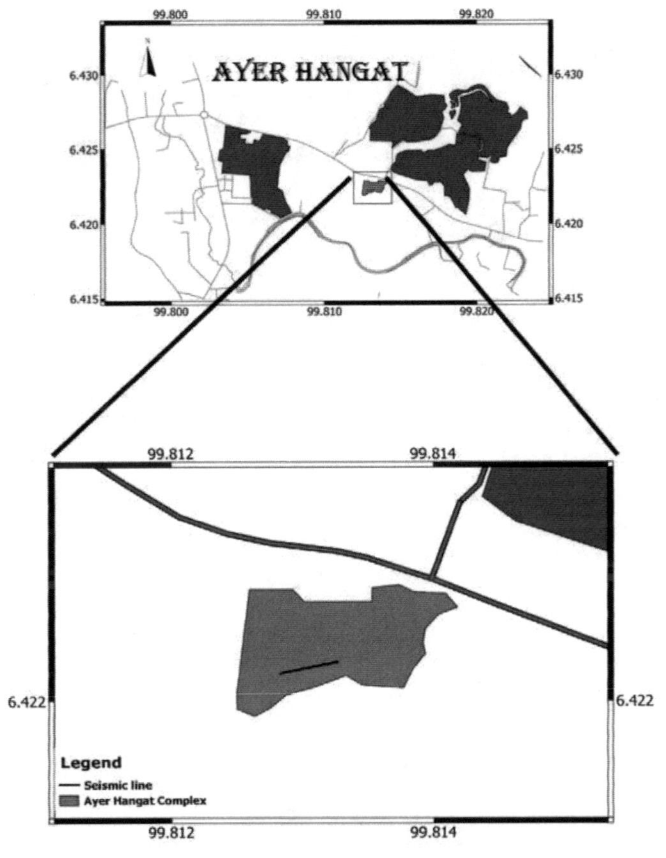

Figure 3. Location map of Seismic line at Ayer Hangat complex.

3.2 Data Processing

From the previous section we realize that two problems are encountered with the data files obtained. The DC-shift is crucial and may lead to false reflection on the final seismic section. OBSPY is used to solve recorded data problems. Python script is developed using OBSPY [7] methods to remove both DC-shift and to convert the data into SEG-Y format. Fig. 4 shows shot gather plot after correction of data using OBSPY. At this point common offset gather at 10, 20 and 26 m are investigated in order choose the best offset to use. The common offset gather at 26 m shows the best resolution and hence will be used for further analysis. Seismic UNIX (e.g. [8] and [9]) is then used to process the data. Starting by resampling the data to 0.008 sec in order to reduce processing time. Static corrections

IOP Conf. Series: Earth and Environmental Science **29** (2016) 012023 doi:10.1088/1755-1315/29/1/012023

which account for the change in the depth of the layer just below the surface and is then removed by aligning first arrival. Pullan et al. [1] pointed out that data processing is rather a cosmetic process to make the seismic more readable. Hence the second step is to adjust the gain using AGC for amplitude balancing of the seismic traces. Taper is then applied to reduce the amplitudes of both the earlier arrivals and arrivals near the end of the records that are not reflection data. Digital band-pass filter is then applied to remove unwanted signals or noise. We used band-pass filter to filter out the data outside the band from 10 to 50 Hz. The aforementioned steps are the same proposed by [1] and [2]. We applied more processing steps to remove high frequency noise and to migrate the final seismic section (Figure 5).

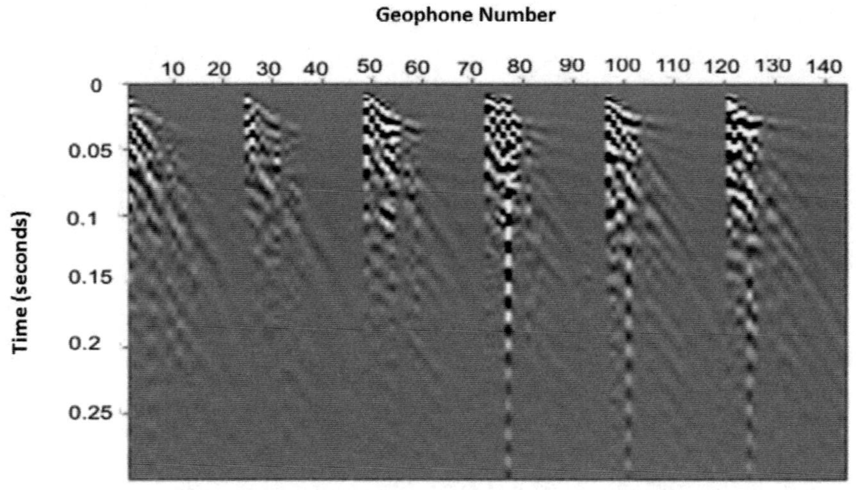

Figure 4. Sample shot gather at Ayer Hangat

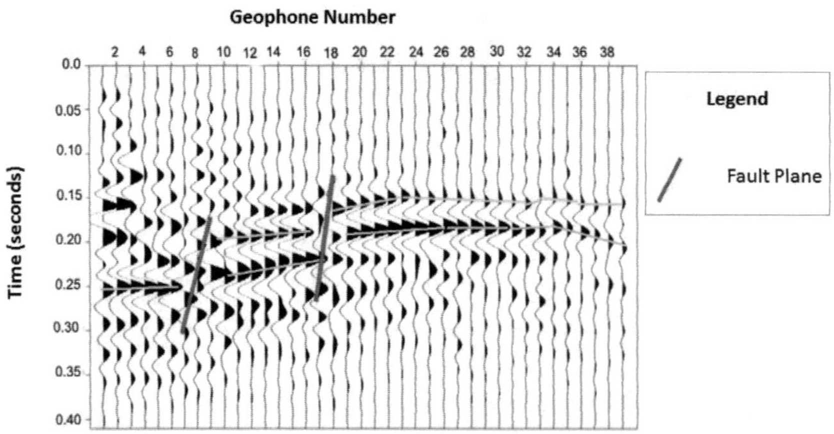

Figure 5. Final model of common offset cross section at Ayer Hangat

4. Discussion

Common offset technique used in the present work aimed towards identifying the near surface structure that may control hot water flow in to the surface. The final seismic section shown in Figure 5 shows that the area near the hot spring is dissected by step faults that apparently dipping in the direction of hot spring (Figure 6). This may indicate that these step faults are controlling hot water flow into the surface. Current analysis using common offset prove successful in determining the structures at near two way times (about 0.2 sec). Depth determination is not included in the present work due to some limitations in the data acquired. However, the deduced result is useful for pointing out where to drill for the development of the geothermal complex.

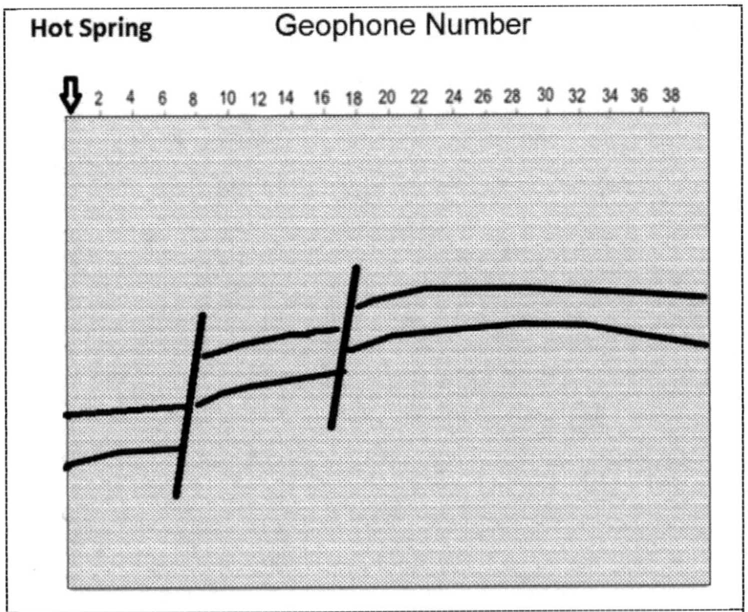

Figure 6. Sketch showing the interpreted subsurface structure.

5. Conclusion

Near surface structures is investigated using common offset seismic reflection technique in order to help define hot water accumulation and pathway to the surface. Forty eight shot points were used at horizontal offset of 1 m. Reflection data were recorded using TERRALOC MK8 instrument with land streamer of 24 geophones at 1 m interval. The record length is 0.8192 sec with sampling interval of 100 us. Data are processed and analyzed using both OBSPY and Seismic Unix software. The final section obtained using common offset of 26 m shows the existence of step faults that dip in the direction of the hot spring. These faults may control the flow of the hot water to the surface. The results of the present study may help in the development plan of the hot spring complex.

References

[1] Pullan S E, Miller R D, Hunter J A, & Steeples D W 1991 Shallow Seismic Reflection Surveys: CDP or "Optimum Offset" SEG Annual Meeting Society of Exploration Geophysicists

[2] Pullan S E, & Hunter J A 1990 Delineation of buried bedrock valleys using the optimum offset shallow seismic reflection technique *Geotechnical and environmental geophysics* **3** 75-87

[3] Whiteley R J, Hunter J A, Pullan S E, & Nutalaya P 1998 "Optimum offset" seismic reflection mapping of shallow aquifers near Bangkok, Thailand *Geophysics* **63** (4) 1385-1394.

[4] Leman M S, Komoo I, Mohamed K R, Ali C A, Unjah T, Othman K, and Yasin M H 2008 Geology and Geoheritage Conservation Within Langkawi Geopark, Malaysia Global Network of National Geoparks (GGN)

[5] Jones C R 197 The geology and mineral resources of Perlis, north Kedah and the Langkawi Islands (Geological Survey Headquarters, Kuala Lumpur)

[6] Ali C A 2008 Field guide to geology of Langkawi Geopark (Institute Alam Sekitar dan Pembangunan (LESTARI), Bangi, Selangor)

[7] Beyreuther M, Barsch R, Krischer L, Megies T, Behr Y, and Wassermann J 2010 ObsPy: A Python Toolbox for Seismology Seismological Research Letters **81** (3) 530-533
[8] Templeton M E, Gough C A 1998 Web Seismic Un*x: Making seismic reflection processing more accessible, Computers and Geosciences
[9] Stockwell Jr J W 1997 Free Software in Education: A case study of CWP/SU: Seismic Un*x The Leading Edge

IOP Conf. Series: Earth and Environmental Science **29** (2016) 012024 doi:10.1088/1755-1315/29/1/012024

Static Electrode DC Resistivity Measurement at Surface Water for Pond Subsurface Layer Imaging

Sumintadireja P[1] and Irawan D[2]
[1]Applied Geology Research Group, Faculty of Earth Science and Technology – Institut Teknologi Bandung.
[2]Graduate Program in Geophysical Engineering, Faculty of Mining and Petroleum Engineering – Institut Teknologi Bandung.

E-mail: prihadi@gc.itb.ac.id

Abstract. Resistivity methods in marine applications show slightly different processing techniques from land based resistivity surveys. Special DC resistivity instruments need to overcome difficulties in arranging the electrode with straight array lines and position. Some geoelectrical instrument manufacturers developed equipment which is able to measure resistivity values and positions in real time. In this paper we demonstrate an application of ordinary geoelectrical instruments for resistivity acquisition in water environment. This study is motivated by the inability to apply conventional DC resistivity instruments in water environment. Land resistivity survey array is arranged on the surface of water using static electrode mode. The method has been tested in various environments, such as ponds/lakes with quiet until rough waves and also measurements at coastal environments. Measurement at the ponds/lakes water environment resulted in data that are almost identic to the measurements obtained using standard land DC resistivity method. On the other hand the measurement in coastal environment does not work properly, possibly due to the lack of power source.

1. Introduction

Geo-electrical measurement uses a multi-electrode system and 2D resistivity inverse modelling can infer the subsurface resistivity structure well. Not only measurement at the ground can be done but also in the water. Electrical resistivity investigations applied in marine environments can be found since the mid-1930s [1] and mid-1980s [2] Since then, applications for resistivity in marine environmental have spanned a wide array of problems that mostly mimic terrestrial applications, but with a few that are unique to the marine setting [3]. Special DC resistivity instruments need to overcome difficulties in arranging the electrode with straight array lines and position. Several commercial instruments are available for acquiring DC resistivity for marine, which primarily need the ability to automatically log voltage data for a given electrode array which is able to measure resistivity values and positions in real time. Most modern equipment also has the ability to accept GPS data to help georeference each measurement. Supported by these advanced tools some experts have research ad about geological mapping using marine resistivity data (e.g. [4] and [5]) and also about submarine groundwater discharge ([6] to [9]).

[1] To whom any correspondence should be addressed.

This paper describes the application of conventional geoelectrical measurements, which usually are applied to the land, for resistivity data acquisition in water environment. However the limitation of this method is that it must be ensured that the electrode is in static mode.

2. Method

There are two kinds of methods to arrange the electrodes in the static mode. First, the acquisition of resistivity in aquatic environments can be conducted at the water surface with floating electrodes (e.g. [10] to [12]) but the tip of each electrode needs to tethered to fixed floating electrodes (Figure 1a). Second, placed submerged at the floor (Figure 1b), as has been done by Toran et al. [13]. Both strategies are typically used in conjunction with 2D subsurface layer imaging using DC resistivity conventional instrument.

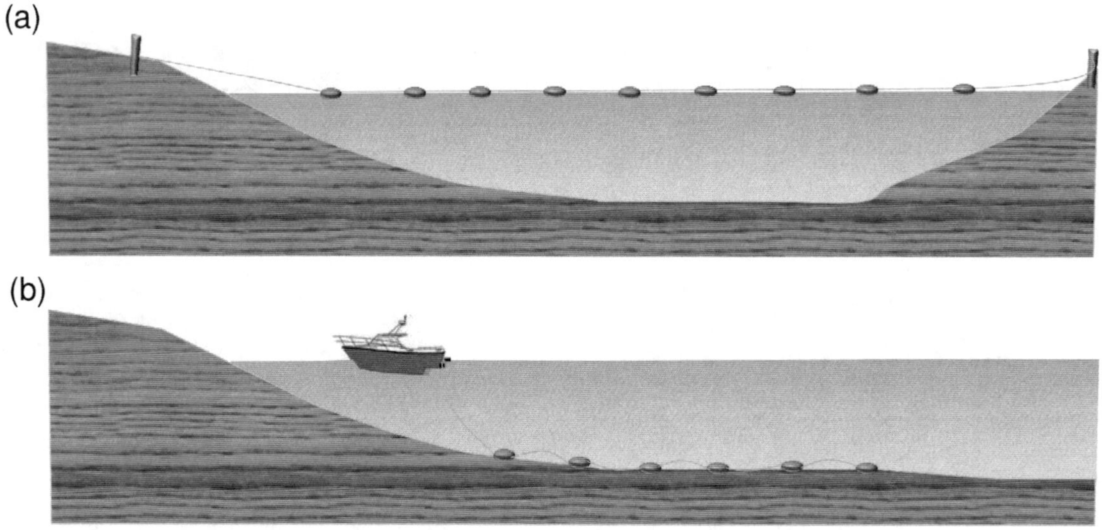

Figure 1. Two techniques to arrange electrodes in static mode

In this study, we use the first method to measure 2D DC resistivity at surface water for pond subsurface layer imaging in the artificial ponds at Eastern Bandung, West Java, Indonesia. The pond is located in the vicinity of housing residents. More than 3000 local households liquid waste empty in this pond.

The survey is planned to follow a line as designed Figure 2. There is overlap between land and water in the measurement to see the response / model between water and soil characters. Fixed floating electrodes shown in Figure 3. 48 electrodes with the smallest electrode spacing of 1.5 meters (total length 70.5 m) were used to perform measurements of 2D resistivity data. Twenty two electrodes placed in the water (floating) and another 26 electrodes at the land. We perform measurements in three configuration arrays (Wenner alfa, Wenner-Schlumberger and dipole-dipole [14]) in the same electrode position to get more convincing data.

The data was processed with two-dimensional inversion by RES2DINV [14] to determine conductive or resistive bodies along the line. At the time of modelling there are no constraints, such as the bathymetry and conductivity of the water, in the inversion process to help resolve issues at the water–rock interface. The parameters used in the inversion are the same as with measurement in the land.

Figure 2. 2D DC resistivity line overlap pond

Figure 3. Photos of fixed floating electrodes.

3. Result and Discussion

Figure 4 shows the inversion results of three different configuration arrays. Figure 4(a) is a model of the measurement using Wenner-alfa, Figure 4(b) from Wenner-Schlumberger and Figure 4(c) from dipole-dipole configuration. Every model was obtained from 5 iterations with rms error not more than 3%. Generally the three models look relatively similar in that the body anomalies have similar depth and thickness.

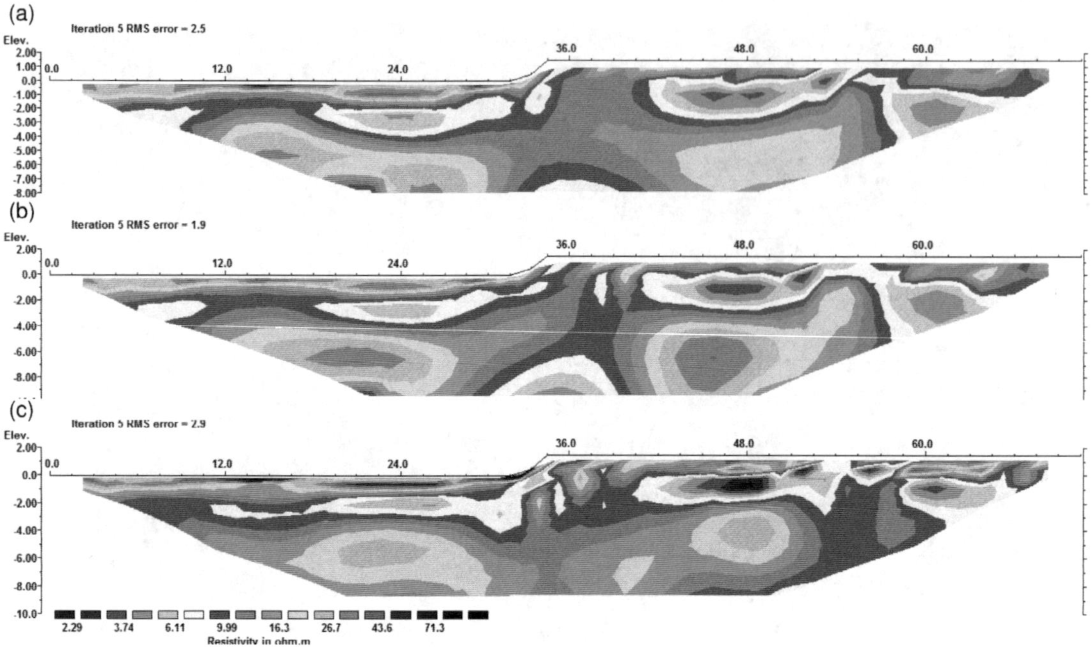

Figure 4. Models resistivity with topography for each electrode array. (a) Wenner Alfa, (b) Wenner-Schlumberger, (c) dipole-dipole configuration

Geological model interpretation was built using history-guided approach. Figure 5a shows that at the time first digging, the pond has been extracted 5 meter deep, more than 14 years ago (excavated in 2001). Next, Figure 5b, more than 15 year since the ponds has been flooded until now. The main problem of this pond is about sedimentation. Not too long after first excavation the pond has been already filled by sedimentation to a thickness of approximately 3 meters. The sedimentation came from downstream irrigation around area (Figure 5c). More than 3000 local houses empty liquid waste to the pond. It makes the water of pond more alkaline with a relatively high pH value. It is one of the reasons why the water is very resistive. On the other hand, in 2013 the local residents dug a well for getting fresh water. The location of well is at a distance of 48 meters from the DC resistivity line. The existence of fresh water made the surrounding conventional well area more conductive. The recent condition of geological model then is shown in Figure 5d. The final geological model shows good correlation with the inversion model result look.

The same method has been tested in various environments, such as ponds/lakes with quiet until rough waves and also measurements at coastal environments. The result has produced good imaging of sub-surface under area target. Unfortunately a measurement at the coastal environments does not work properly, possibly due to the lack of power source. In highly conductive seawater, higher powered systems (>1kW) are recommended to overcome the noise. Corwin et al. [2] state that 10 A of current was sufficient to overcome noise in seawater for a submerged Schlumberger array with electrode spacing maximum (AB/2) = 316 m.

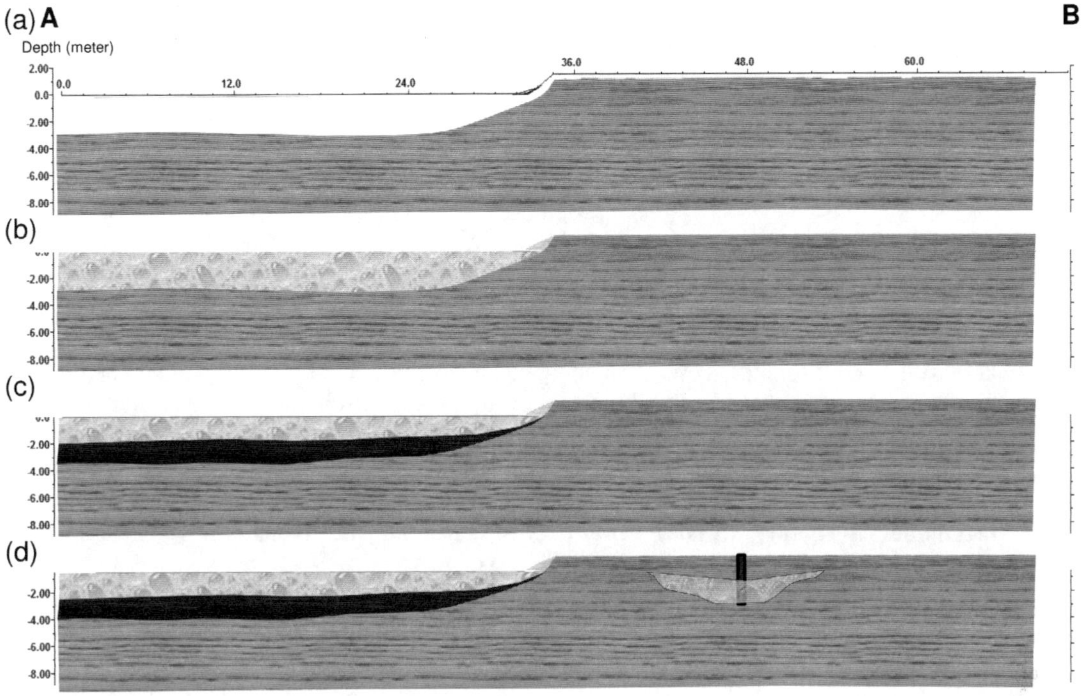

Figure 5. Geological model from history of pond. (a) Pond is extracted will 5 meter depth, (b) flooded for 15 years, (c) sedimentation from downstream irrigation around area, (d) local residents dug well (in 2013)

4. Conclusion

Static electrode DC resistivity measurement at surface water at artificial ponds has produced good imaging relation with geological model using historical approach. Measurement at the ponds/lakes water environment provided data that are almost identic to the measurements obtained using standard land DC resistivity method using low power instruments. For freshwater environments, where the resistivity values of the water column are relatively high, low powered systems (200–400 W) can be used quite effectively.

References

[1] Schlumberger, C., Schlumberger, M., Leonardon, E.G., 1934. Electrical exploration of water-covered areas. Transactions of the American Institute of Mining and Metallurgical Engineers 110, 122–134.

[2] Corwin, R.F., Turnross, J.L., Harding Jr., R.S., 1985. Offshore applications of the directcurrent resistivity method. SEG Expanded Abstracts 4 (1), 230–231.

[3] Loke, M.H., Chambers, J.E., Rucker, D.F., Kuras, O. and Wilkinson, P.B., 2013. Recent developments in the direct-current geoelectrical imaging method. Journal of Applied Geophysics 95 135–156.

[4] Rinaldi, V., Guichon, M., Ferrero, V., Serrano, C., Ponti, N., 2006. Resistivity survey of the subsurface conditions in the estuary of the Rio de la Plata. Journal of Geotechnical and Geoenvironmental Engineering 132 (1), 72–79.

[5] Rucker, D.F., Noonan, G., Greenwood, W.J., 2011. Electrical resistivity in support of geological mapping along the Panama Canal. Engineering Geology 117 (1–2), 121–133.

[6] Day-Lewis, F.D., White, E.A., Johnson, C.D., Lane Jr., J.W., 2006. Continuous resistivity profiling to delineate submarine groundwater discharge–examples and limitations. The Leading Edge 25 (6), 724−728.

[7] Henderson, R.D., Day-Lewis, F.D., Abarca, E., Harvey, C.F., Karam, H.N., Liu, L., Lane, J.W.J., 2010. Marine electrical resistivity imaging of submarine groundwater discharge: sensitivity analysis and application in Waquoit Bay, Massachusetts, USA. Hydrogeology Journal 18 (1), 173−185.

[8] Swarzenski, P.W., Burnett, W.C., Greenwood, W.J., Herut, B., Peterson, R., Dimova, N., Shalem, Y., Yechieli, Y., Weinstein, Y., 2006. Combined time-series resistivity and geochemical tracer techniques to examine submarine groundwater discharge at Dor Beach, Israel. Geophysical Research Letters 33, L24405.

[9] Swarzenski, P.W., Simonds, F.W., Paulson, A.J., Kruse, S., Reich, C., 2007. Geochemical and geophysical examination of submarine groundwater discharge and associated nutrient loading estimates into lynch cove, Hood Canal, WA. Environmental Science & Technology 41 (20), 7022−7029.

[10] Hatch, M., Munday, T., Heinson, G., 2010. A comparative study of in-river geophysical techniques to define variation of riverbed salt and aid managing river salinization. Geophysics 75 (4), WA135−WA147.

[11] Kelly, B.F.J., Allen, D., Ye, K., Dahlin, T., 2009. Continuous electrical imaging for mapping aquifer recharge along reaches of the Namoi River in Australia. Near Surface Geophysics 7 (4), 259−270.

[12] Song, S.H., Cho, I.K., 2009. Application of a streamer resistivity survey in a shallow brackish-water reservoir. Exploration Geophysics 40 (2), 206−213.

[13] Toran, L., Johnson, M., Nyquist, J., Rosenberry, D., 2010. Delineating a road-salt plume in lakebed sediments using electrical resistivity, piezometers, and seepage meters at Mirror Lake, New Hampshire, U.S.A. Geophysics 75 (4), WA75−WA83

[14] Loke, M.H., 2012. Tutorial: 2-D and 3-D Electrical Imaging Surveys. Geotomo Software, Malaysia.

Influence of Lithology and Slope Gradient to Infiltration of the Mount Malabar, West Java

A Pratama , D E Irawan, A Susanto, and R D W Ardi
Dept of Geology, Faculty of Earth Sciences and Technology,
Institut Teknologi Bandung, Jalan Ganesa No. 10, Bandung 40132, Indonesia

E-mail: aditya16311209@yahoo.co.id

Abstract. Volcano is an area which serves as a catchment area for the lowlands. Ability of rock or weathered-soil to absorb the rain water depends on several things, such as lithology and large of slope. Different lithology has different characteristics, including in terms of porosity which is directly related to the ability of rock to store water. Characteristics of lithology in volcanic area can change rapidly, both vertically and laterally. Large of slope in volcanic area that change significantly also can affect the infiltration rate (the seepage of rain) in rock or weathered-soil. Therefore, the influence of lithology and large of slope to the infiltration rate should be proven to predict the infiltration zone in volcanic area. Observations has been conducted on the eastern slopes of Mount Malabar with an area 78 km^2, at coordinates $7^0$03'28.04" LS - $7^0$10'32.05" LS and $107^0$38'37.64" BT - $107^0$41'50.6" BT. The infiltration rate observed on the weathered-soil using simple single infiltrometer made of PVC pipe 50 cm long, on March-April 2015. The measurement is carried out at several points where the weathered-rock result has been known, as much two times for different slope in each point. 26 measurement points have been obtained from different slopes and weathered-soil of different five-lithology. The results showed that the infiltration rate proportional to the percentage of rock porosity and large of slope. Infiltration rate sequence from the smallest to the greatest are weathered-soil andesites, basaltic andesite, laharic breccias, alteration of dacite, and pyroclastic breccias. The greatest infiltration rate obtained is 10.11 cm/minute in pyroclastic breccia with 25° slope, while the smallest is 0.0437 cm/minute in pyroclastic breccias with 4° slope.

1. Introduction

Volcano territory is one of areas that have many benefits. In addition of being recognized with the abundance of mining and energy resources including geothermal, volcano territory also has the potential to become a recharge zone. Mount Malabar is one of a type B volcanoes located in South Bandung, West Java which has natural resources. Mount Malabar is one of catchment area [1]. However, land clearing for plantations and agriculture cause its function as catchment area to be disturbed. The existence of the plant at ground level can impede the flow of water on the surface thus the infiltration becomes greater. In addition, its roots can speed mechanical weathering into soil [2]. The thicker the soil, the larger infiltration capacity [2]. Infiltration is one of the parameters to determine the catchment area potential. Potential of Mount malabar as catchment area must be analyzed in order to keep its function. Along with the increasing of water that infiltrating the soil, the run off water volume that considered as one of causes of flood is decreasing. The infiltration rate affected by many factors including soil texture and slope gradient [3]. This research will focus on those two due to the differences in geology and geomorphology condition between volcano territory and plain territory. According to [4] the key characters of volcano territory is the rapid lithological change both vertically and horizontally, in addition the various morphology configuration.Therefore it will be best to know how lithology and geomorphology condition affected infiltration rate in the research area. Results from this research are expected to be a reference in determining the catchment areas potential in future research.

2. Theory

In a hydrogeology cycle the rain drops that reach earth surface will be undergoing 2 process which is flow on the surface as run off and infiltrating underground. The infiltration process of rain water to unsaturated known as infiltration [3]. Meanwhile, capacity to absorb the rain water known as infiltration capacity [5]. The iniltration ability is different in any weathering result soil. This condition affected by 5 factors including precipitation characteristic,earth surface condition, earth surface cover condition, ground texture and structure, and also the infiltrated water characterictics [3]. According [5] the infilttration capacity for any soil also different that can be approached by analyze the soil porosity or the origin rock that form the soil. In addition of being a container for fluid, porosity also relate with permeabiltiy which is the main factor of fluid movements inside of soil or rock. According to[6] porosity is lineary linked with permeability logarhtym, but not in several case due to rock/soil textures such as sorting and grain size. Soil texture can determine whether the porosity in rock/soil interconnected or not where the interconnected porosity will will increase the permeability. In addition the high porosity and permeability make the weathering more intensive therefore the weathering result soil will formed faster.

3. Research Method

The infiltration rate observed on the weathered-soil using simple single infiltrometer made of PVC pipe 50 cm long and 8 cm diameter, on March-April 2015. That pipe plugged into the soil with 10 cm in depth and then fill it with water until 40 cm of height. Then we measure the water height in 0 minutes, 1 minutes, 2 minutes, 3 minutes and so forth until the water height relatively constant and not decreasing further (Figure 1). After the measurement done then we measure the slope gradient of this location using geological compass and board as pedestal. The measurement is carried out at several points where the weathered-rock result has been known, as much two times for different slope in each point. Ten measurement points have been taken for each weathered soil from 4 lithology also for each lithology the measurement conducted for several different slope gradient. Finally we got the

Figure 1. Simple Single Infiltrometer Device

infiltration rate per minute and then we process it with Microsoft Excel based on the equation (1) [7]:

$$f = a\,t^{n} \tag{1}$$

f = infiltration rate; a, n = constanta; t = time (minutes)

"a" and "n" constanta were evaluated from cumulative infiltration rate for determined times. The integral form of cumulative infiltration rate shown in the equation (2) :

$$F = \int_0^t a\,t^{n} = \frac{a}{n+1} t^{n+1} \tag{2}$$

F = Cumulative Infiltration Rate

4. Regional Geology

This Research located in Malabar Volcanoes area which which in administrative belong to Banjaran and Arjasari District, Bandung Regency, West Java Province. Geographycally the research area located at 7° 03' 42" - 7° 10' 44" LS dan 107° 38' 53" - 107° 41' 36" BT. Research area shown in red rectangle and cover 78 km^2 (13 x 6 km) area which covered the east slope area of Malabar Volcanoes. According to [8] Southern Bandung Area can be divided into eleven unit, they are: Soreang VolcanicRock Unit (SV), Baleendah Volcanic Rock Unit (BV), Pengalengan Volcanic Rock Unit (PV), Tanjaknangsi Volcanic Rock Unit (TV), Kuda Volcanic Rock Unit (KV), Kendeng Volcanic Rock Unit (KdV), Dogdog Volcanic Rock Unit (DV), Wayang-Windu Volcanic Rock (WV), Malabar Volcanic Rock Unit (MV), Pengalengan Pyroclastic Unit, and Alluvium Unit (Figure 2A).

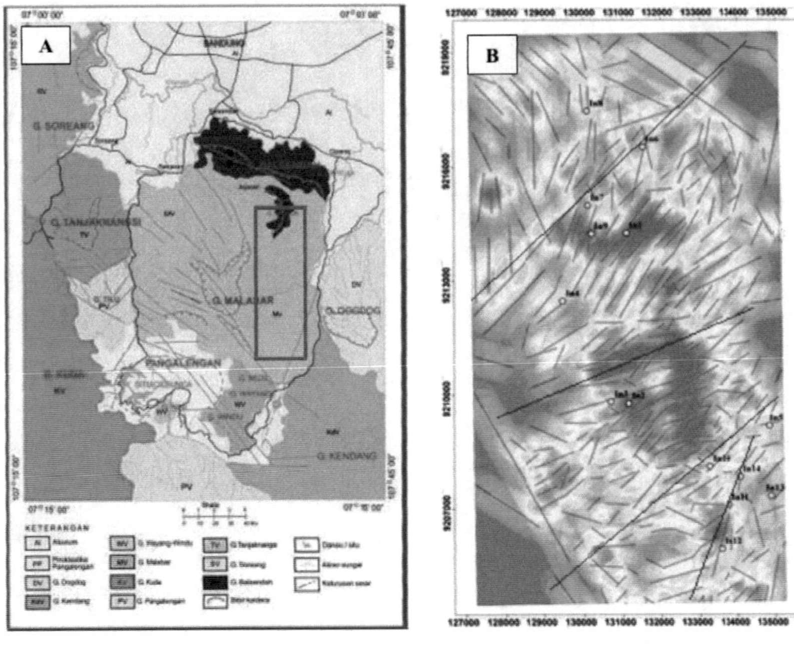

Figure 2. A) Geological Map of South Bandung [1], B) Fault and Fracture Density (FFD) map of research area, yellow circle show measured point of infiltration rate, red line show lineaments, blue line show fault interpretation

Based on this classification the research area belong to Malabar Volcanic Rock Unit and Baleendah Volcanic Rock Unit which is consist of basaltic – andesit-basaltic lava, pyroclastic breccia, laharic breccia, tuff, Baleendah layered lava flow with pyclastics breccia infix.

According to [8] by analyzing landsat image can be determined 2 main lineation pattern in South Bandung Region which is Northwest-Southeast and Northeast-Southwest. This statement match with landsat image observation result in research area. Lineament analysis executed with SRTM image and then converted to Fault and Fracture Density (FFD) in ArcGIS software (Figure 2B). FFD is one of methods used to determine fault and fracture density in area [9]. Fracture formed by the process of deformation due to the tectonic activity. That picture show 2 main lineation pattern in resaearch area which is match with regional lineation pattern in South Bandung Region. In addition, from the continuation of lineation pattern can be interpreted 4 major fault in reseach area with Northeast-Southwest relative orientation.

5. Discussions and Results

Infiltration rate measurement executed in 14 different location with 26 total measurement. Field observation show 5 different lithology which is the origin rock for wearhering result soil in the measurement location (Figure 3). Those lithology are pyroclastics breccia, laharic breccia, altered dacite, andecite, and basaltic-andecite with characteristics given below:

Figure 3. Outcrop image and hand specimen of : A) Pyroclastic Breccia, B) Laharic Breccia, C) Altered Dacite, D) Andecite basatic, E) Andecite

1. Pyroclastic breccia; Yellowish brown-dark brown; grain size pebble-boulder (Wentworth,1922); grain shape angular; poor sorted; matrix supported; fragment including lithic andecite, lithic andecite basaltic, and lithic dacite; matrix ash-lapili including lithic and volcanic glass.
2. Laharic breccia; yellowish brown-dark brown; grain size pebble-boulder (Wentworth,1922); grain shape subangular-angular; very poor sorted; matrix supported; fragment including lithic andecite, lithic andecite basaltic, and lithic dacite; matrix ash-lapili including lithic and volcanic glass.
3. Dacite; bluish gray; aphanitic, porphyritic, inequigranular, holocrystaline, hypidiomorphic, mineral consist of plagioclase (30%) and clay mineral (70%)
4. Andecite; light gray, aphanitic, porphyritic, inequigranular, holocrystaline, hypidiomorphic; mineral consist of plagioclase (40%), quartz (5%), hornblende (10%), aphanitic minerals (45%).
5. Andecite basaltic; dark gray; aphanitic; porphyritic; inequigranular; holocrystaline; hypidiomorphic; mineral consist of plagioclase (15%), and aphanitic mineral (85%).

6. Lithology vs Infiltration Rate

The 26 numbers of infiltration rate measurement data classified into 5 groups according to the rock origin of the soil. In each group the average infiltration rate calculated. From that calculation we get that the infiltration rate from highest to lowest is in origin rock pyroclastic breccia, altered dacite, laharci breccia, andecite basaltic, and andecite (Figure 4). This result match with the porosity of each rocks. According to petrographic analysis the porosity percentage from highest to lowest is pyroclastic breccia matrix (33%), altered dacite (30%), laharic breccia matrix (23%), andecite basaltic (7%), and andecite (5%). The porosity in thin section pointed by blue dyed area (Figure 5). The high porosity made a high infiltration capacity that push the formation of secondary porosity due to dissolution because the infiltrating water can easily reacted with volcanic material that formed the rock which increase the permeability.

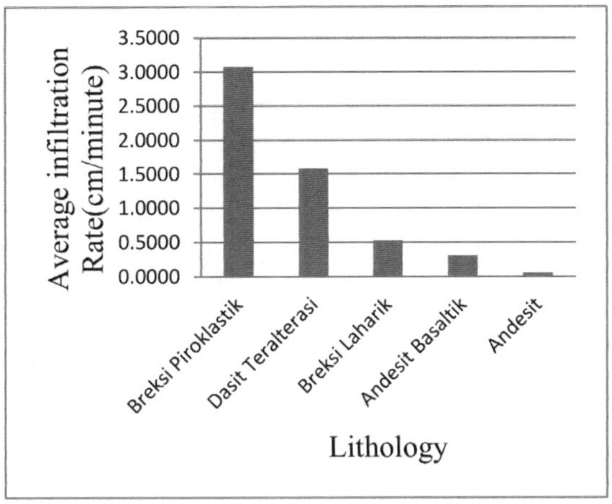

Figure 4. Lithology vs Infiltration Rate Graphic

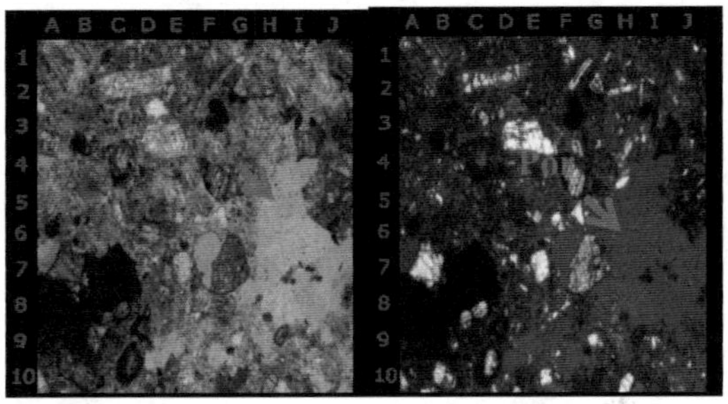

Paralel Nicol
Lensa perbesaran 2.5 x
Code : 203
Rock name : Matrix of Laharic Breccia
Porosity : 25%

Cross Nicol
Lensa perbesaran 2.5 x
Code : 203
Rock name : Matrix of Laharic Breccia
Porosity : 25%

Figure 5. Thin section of Laharic Breccia

The high porosity and permeability cause the high infiltration rate that make weathering process more intensive. This statement support by the fact that thesoil from pyclastic breccia weathering can reach 10-25 m thick meanwhile the soil from the andecite weathering only reach 1-2 m thick. The high porosity and intensive weathering in dacite is related to the alterion process that happened. According to [12] rock that experienced hydrothermal alteration will form secondary mineral and make rock softer also has more porosity. In addition the thicker soil from andecite basaltic weathering can be affected by geological structure. Measurement in red zone which characterize the biggest lineation density and located near the interpreted fault zone.

Fracture which formed because of fault related structure can become a secondary porosity and become the infiltration way for surface water to the rock, therefore make the weathering process faster. This argument match with [10] statement which state that volcanic rocks is a good permeabel aquifer, one reason beacuse of the effect from faulting and cracking that formed after the rock formation.

7. Slope Gradient vs Infiltration Rate

The Slope Gradient vs Infiltration Rate Graphic only made in 2 groups, they are origin rock pyroclastics breccia and laharic breccia because statistically number of measurements in both groups fulfilled the criteria, 14 data for pyroclastic breccia and 5 data for laharic breccia. Meanwhile the number of measurement data for altered dacite, andecite, and andecite basaltic just two for each group. Koefisien korelasi pada kedua grafik tersebut menunjukan nilai positif, yaitu 0.0279 dan 0.0171 (Figure 6). The coefficient correlation can be used to determine the relationship between two variables, in this case the slope and the infiltration rate. According to [11], two variables that are interconnected indicated with R^2 close to 1 or -1. If R^2 is positive, the two variables are linearly related. If the value of the variable increased by "x" then the other variables increased by "x". Whereas, when R^2 is negative, the two variables are inversely related. If the value of the variable increased by "x" then the other variables decreased by "x". The result is in pyroclastic breccia and laharic breccia slope gradient is weak linearly related with infiltration rate. According to [12] this condition caused by the relatively looser soil particle in higher slope gradient therefore water will be easier to infiltrate. Meanwhile in lower slope gradient the soil is more compact. In this case soil texture become the main factor that control the infiltration rate. From field observation we know that in steeper slope the soil particle is looser and not compact. The compaction that affected by human and animal activity also considered one factor that made soil in low slope gradient more compact.

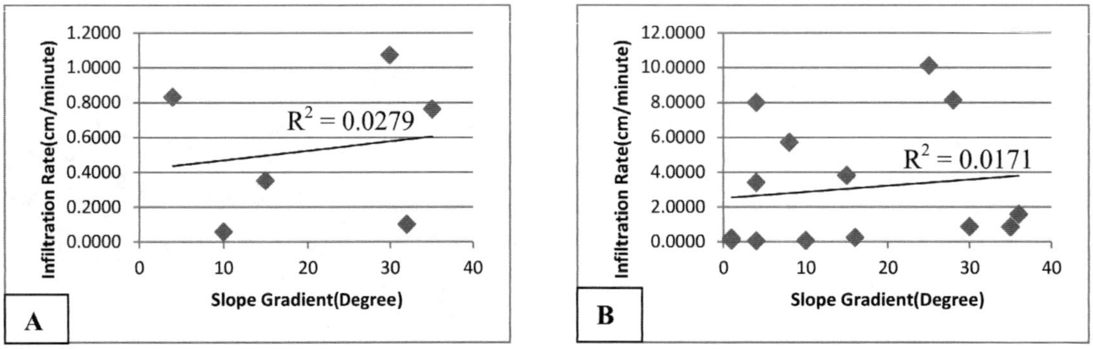

Figure 6. A) . Slope Gradient vs Infiltration Rate Graphic in laharic breccia origin rock, B) Slope Gradient vs Infiltration Rate Graphic in pyroclastic breccia origin rock

8. Conclusion

Infiltration rate proportional to the percentage of rock porosity and large of slope. Infiltration rate sequence from the smallest to the greatest are weathered-soil andesites, basaltic andesite, laharic breccias, alteration of dacite, and pyroclastic breccias.

The greatest infiltration rate obtained is 10.11 cm/minute in pyroclastic breccia with 25° slope, while the smallest is 0.0437 cm/minute in pyroclastic breccias with 4° slope. The high infiltration rate in pyroclastic rock relates to the secondary porosity forming due to dissolution which increase permeability. Hydrothermal alteration process cause rock to have a good porosity and permeability which make the high infiltration rate. Secondary porosity (fracture) in andecite basaltic become the infiltration way which cause more intensive weathering. The high slope gradient cause the soil became looser and uncompact therefore make the water easier to infiltrate. Hi-res and full size images are available on the following link https://onlinewaterbook.wordpress.com/2015/07/02/supplementary-page-influence-of-lithology-and-slope-gradient-to-infiltration-of-the-mount-malabar-west-java/.

9. References

[1] Narulita I, Rahmat A and Maria R 2008 Aplikasi Sistem Informasi Geografi untuk Menentukan Daerah Prioritas Rehabilitasi di Cekungan Bandung *J.Riset Geologi dan Pertambangan Jilid 18.* **1** 23-35
[2] Harto S B 1993 *Analisis Hidrologi* (Jakarta: Gramedia Pustaka Utama)
[3] Wibowo H 2010 Laju Infiltrasi pada Lahan Gambut yang Dipengaruhi Air Tanah (Study Kasus Sei Raya Dalam Kecamatan Sei Raya Kabupaten Kubu Raya) *J.Belian* pp 90-103
[4] Yuwono Y S 2004 *Pemetaan Daerah Volkanik* (Bandung: Insitut Teknologi Bandung)
[5] Irawan D E and Puradimaja D J 2015 *Hidrogeologi Umum* (Bandung: Penerbit Ombak)
[6] Nurwidyanto, M I, Noviyanti I and Widodo S 2005 Estimasi Hubungan Permeabilitas dan Porositas Pada Batupasir (Studi Kasus Formasi Kerek,Ledok, Selorejo) *Berkala Fisika* pp 87-90

[7] Dhalhar M A 1972 Process and Field Evaluation of Infiltration Rate. (Minneapolis: University of Minnesota)

[8] Bronto S, Koswara A, and Lumbanbatu K 2006 Stratigrafi Gunungapi Daerah Bandung Selatan, Jawa Barat *J. Geologi Indonesia* 89-101

[9] Suryantini and Wibowo H H 2010 Application of Fault and Fracture Density (FFD) Method for Geothermal Exploration in Non-Volcanic Geothermal *J. Geoaplika.* **5** 27-37

[10] Todd D K. and Mays L W 2005 *Groundwater Hydrology* (New York: John willey & Sons. Inc)

[11] Lattin J, Carrol J D and Green P E 2003 *Analyzing Multivariate Data* (Toronto: Curt Hinrichs)

[12] Budiman J S 2012 *Geologi dan Studi Infiltrasi Daerah Lembang Selatan, Kabupaten Bandung Barat, Provinsi Jawa Barat* (Bandung: Institut Teknologi Bandung)

Conceptual model of groundwater and river water interactions in Cikapundung riverbank, Bandung, West Java

Achmad Darul[1], Dasapta Erwin Irawan[2,*], Nurjanna Joko Trilaksono[2], Aditya Pratama[2], Ulfi Rizki Fitria[3]

[1] Dept. of Geology Engineering, Faculty of Engineering and Design, Institut Teknologi dan Sains Bandung, Kompleks Deltamas, Bekasi, Indonesia.
[2] Faculty of Earth and Science Technology, Institut Teknologi Bandung, Jalan Ganesha No. 10 Bandung-40132, Indonesia.
[3] Dept of Mine Exploration Engineering, Faculty of Engineering and Design, Institut Teknologi dan Sains Bandung, Kompleks Deltamas, Bekasi, Indonesia.

E-mail: erwin@gc.itb.ac.id

Abstract. Cikapundung river holds a very strategic role as one of the water supply source for Bandung, but the water quality is decreasing over the years. This degradation has also influenced the groundwater in its riverbank. This paper discusses our effor to build a conceptual model to reconstruct previous analytical model which was built in 1997. The base map was built using a total of 32 scenes of WorldView-2 image, combined with Aster image. The dimension of the model was one layer model with elevation (Z) 400 to 2200 m, 361 rows and 454 columns, covering the area of 21.6 x 27.2 km^2. Two types of property hydrogeology were defined based on the existing geological maps: Volcanic breccias and, Sand and clay intercalations. The parameterization of the stream was divided into two segments: Curug Dago to Viaduct, Viaduct to Dayeuhkolot. Initial head were observed at: 17 dug wells, 24 river points, and two spring points in 2013. All parameters accommodated by software SWS visual MODFLOW flex 2012 that used three-dimension mathematic equation in steady state to build a conceptual model. Model has been calibrated and showed an appropriateness with dug wells and springs with correlation coefficient of 0.92 and water balance in 0.01 steady state condition (2040 m3).The conceptual model successfully replicates the previous analytical model, showing three segments of water interactions: no flow at segment Maribaya to Curug Dago, combination of effluent and influent flow at segment Curug Dago to Viaduct, and influent flow Viaduct to Dayeuhkolot. However, the model shows some local variations that was not spotted in the previous model.

1. Introduction

Water from Cikapundung River is a potential source for water supply in several areas in Bandung. The usage, however, has not been increasing due to the low quality. Lubis (1997) has identified three hydraulic relationship between river water and unconfined groundwater along the river bank, as follows from north to south: isolated stream at Maribaya area (upstream), effluent stream (or gaining stream) at Maribaya to Viaduct (Central area of Bandung) segment, and influent stream (or losing stream) from Viaduct to Dayeuhkolot. Since then, the vast growth of Bandung and the land-use change on upstream area leads to the change of groundwater usage and possibly its hydraulic

* To whom any correspondence should be addressed.

connection with river water. This paper discusses our effort to transform analytical research model on hydrodynamic interaction of groundwater-river water in 1997 into a numerical model. the model discussed in the paper is a numerical version of non-numerical previous result. We try to re-model the 1997 model based on new data and finite difference solution in ModFlow.

The regional geological and hydrogeological background of this paper is based on several previous studies by Sudjatmiko (1972), Silitonga (1973), Sutrisno (1983), Koesoemadinata and Hartono (1981), Priowirjanto and Marsudi (1995), Geyh (1990), Matahelumasi and Wahyudin (2009). A compiled geological map is shown in Fig. 1. This watershed is part of greater Citarum watershed. It consists of:

- Northern volcanic highland at 650-1000 masl, with slope of 3 to 15%. It is composed mostly of breccias and lava.
- Southern Alluvium lowland at 600-650 meter above mean sea level (masl), with slope of 0 to 3%. It is a fertile agricultural area composed of volcanic fan with grain size of clay, sand and pebble.

Faults and joints are found in the area, especially with the existing of Lembang fault in the northern volcanic highland. This east-west oriented fault confines the groundwater flow between the north and south block, based on his water quality measurements.

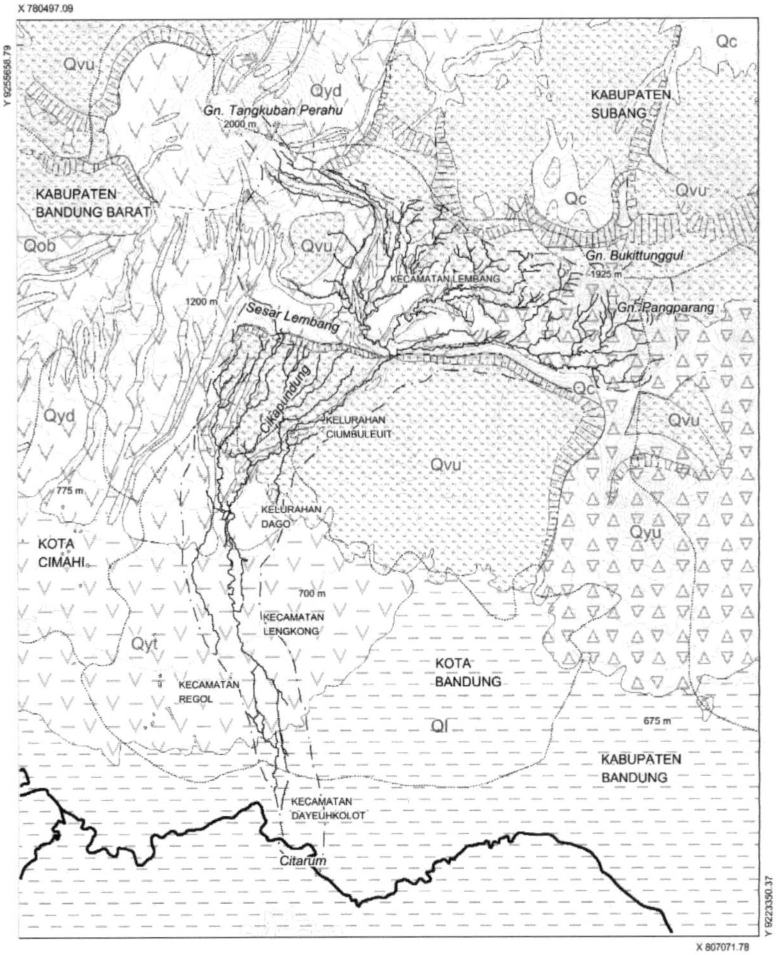

Figure 1. Geology map of Cikapundung watersheed.

2. Methods and Data

The base map was built using a total of 32 scenes of WorldView-2 image, combined with Aster image. The dimension of the model was one layer model with elevation (Z) 400 to 2200 above the sea level, total cell 361 rows and 454 columns, Cell activated was only 144 km^2, as wide as Cikapundung watershed, and the river was set in constant head condition.

Two types of rocks were defined based on the existing geological maps: Volcanic breccias with Conductivity=10^{-6} m/second; Storage=0.02; Specific yield=0.02; Total porosity =0.5; Effective porosity=0.1; Sand and clay intercalations with Conductivity =10^{-8} m/second; Storage =3.3.10^{-6}; Specific yield =0.08; Total porosity =0.11; Effective porosity =0.1. The parameterization of the stream was divided into two segments (width x depth x hydraulic conductivity): Curug Dago to Viaduct 4.5 m x 1.08 m x 10^{-5} m/sec, Viaduct to Dayeuhkolot 4.4 m x 1.5 m x 5.4e10-5 m/sec. Initial head were observed at: 17 dug wells, 24 river points, and two spring points in 2013.

Major problem in setting up the conceptual model was how we determine the boundary conditions, especially in the lateral directions. In this version, we try to adapt the following settings: zero flow as lower and lateral boundary and recharge as upper boundary. The final conceptual model is shown in Fig. 2. The method and preliminary part of the modeling was already published in Darul et al. (2014).

Table 1. Field survey of groundwater level initial head (GW = Groundwater, SW = surface water, SP = Spring)

Name	X	Y	Head (m asl)	T(°C)	DO (ppm)
GW1	787982	9238529	786	26.6	1.2
GW2	788010	9238541	864.92	26.5	5.2
GW3	787003	9238021	770.8	26.9	2.7
GW4	787976	9236921	741.16	25.5	4
GW5	788233	9238797	788.1	23.7	16.21
GW6	788682	9237790	723.61	24.1	7.3
GW7	787672	9237769	721.32	24.2	26.2
GW8	788284	9239120	739.2	23.4	12.2
GW9	788465	9239499	766.65	24.3	17.3
GW10	788985	9239867	857.63	23.8	9.3
GW11	788818	9233207	727.47	24.9	3.4
GW12	788839	9233307	728.54	25.4	14.7
GW13	788680	9233300	720.79	25	12.5
GW14	789027	9233463	718	24.9	14.4
GW15	789665	9230790	699.78	25.3	15.1
GW16	789742	9230799	699.37	24.5	13.9
GW17	789767	9230490	698.2	25.4	17
SW1	788179	9238176	777	23.5	
SW 2	788135	9238101	755	27.8	4.23
SW 3	787923	9237941	760	27.8	5.69
SW 4	787868	9237830	655	24.7	6.32
SW 5	787902	9237567	660	24.7	6.09
SW 6	787997	9237294	769	25	5.88
SW 7	788005	9236998	763	24.7	6
SW 8	787884	9236723	758	24.9	6.31
SW 9	788054	9236484	758	24.1	5.84

Name	X	Y	Head (m asl)	T(°C)	DO (ppm)
SW 10	788026	9236290	763	24.9	5.19
SW 11	788144	9238216	771	23.6	14
SW 12	788014	9238491	781	24.3	24.2
SW 13	788183	9238678	779	28.3	28.3
SW 14	788210	9238842	781	20	20
SW 15	788226	9239041	833	21.4	21.4
SW 16	788000	9235798	734	27.1	21.5
SW 17	788011	9235274	730	24	27.8
SW 18	788184	9235119	729	24.2	24.8
SW 19	788226	9239041	833	23.7	21.8
SW 20	788610	9239508	816	22	11.4
SW 21	788714	9239949	833	22.1	18.1
SW 22	788717	9240010	814	22.7	13.4
SW 23	788727	9233263	729	24.5	14.1
SW 24	789754	9230780	702	24.9	21.9
SP1	788057	9238320	763	24.7	7.5
SP2	787970	9238349	784	24.2	10.1

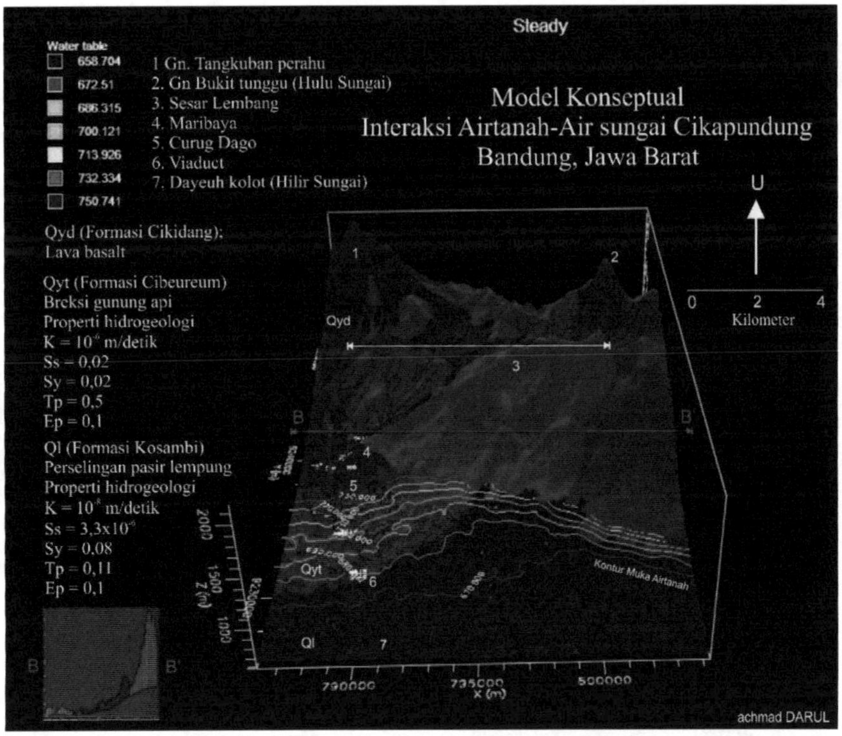

Figure 2. Three dimensions of Cikapundung watershed.

In the field survey, we measured the wells in a distance of 50 m on left and right direction from the river stream used handheld instruments a Solinst water level detector (WLD) for water level depth. This distance is chosen as the max lateral boundary condition. All parameters were loaded in Visual MODFLOW Flex 2012 that used three-dimension finite-difference approximation. The governing 3D equation used in this model was based on MODFLOW (McDonald and Harbaugh, 1988 and Harbaugh, etal., 2000), which combines Darcy's Law and the principle of conservation of mass in steady state condition, as follows.

$$\frac{\partial}{\partial x}\left(K_{xx}\frac{\partial h}{\partial x}\right)+\frac{\partial}{\partial y}\left(K_{yy}\frac{\partial h}{\partial y}\right)+\frac{\partial}{\partial z}\left(K_{zz}\frac{\partial h}{\partial z}\right)+q_s=S_s\frac{\partial h}{\partial t}$$

where K_{xx}, K_{yy}, and K_{zz} are the values of hydraulic conductivity along the x, y, z (major axes of hydraulic conductivity [L/T], H is hydraulic head [L], Qs is volumetric flux of ground water sources and sinks per unit volume [1/T], Ss is specific storage [1/L], and t denotes time [T]

3. Results

Model has been calibrated based on 17 dug wells and two springs with correlation coefficient of 0.92 and water balance in 0.01 steady state condition (2040 m^3) (see Fig. 3 and Fig. 4). We believe the calibration was fairly successful considering the simplification on the geology we had to overcome in the model.

There is no significant difference in water interaction boundaries if compared to 1997 model. The numerical model has also shown three zones (Fig. 5). However, we can see several anomalous spots in the effluent zone (Dago to Viaduct), which were not found in the old map. In such zones, local groundwater drawdown occurs due to possible near by groundwater pumping, presumably deep wells owned by local small-size hotels. This phenomena indicates the nature of multi-aquifer system in the basin, in which, the shallow and deep groundwater are interconnected. Shallow groundwater level will eventually drop along with extensive deep groundwater discharge in the vicinity. With the close relation of shallow groundwater and river water, seepage of river water into the aquifer is possibly inevitable.

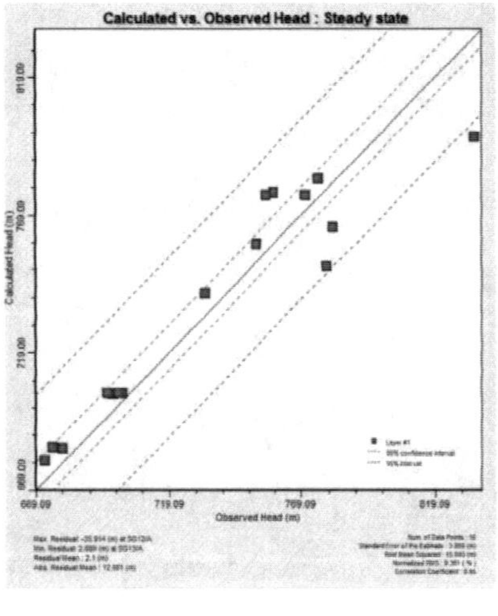

Figure 3. Calibration of head.

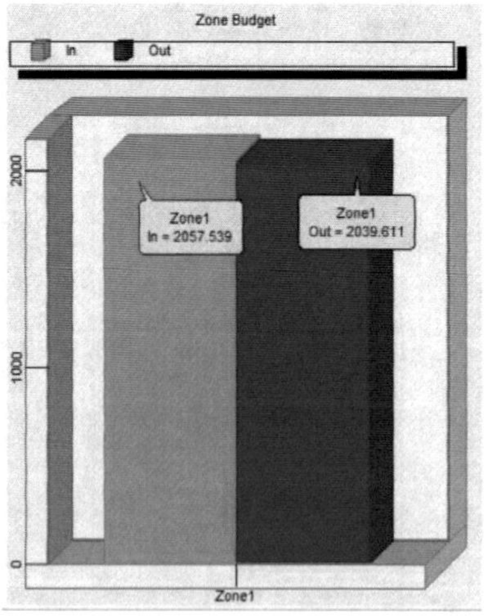

Figure 4. Calibration of water balance.

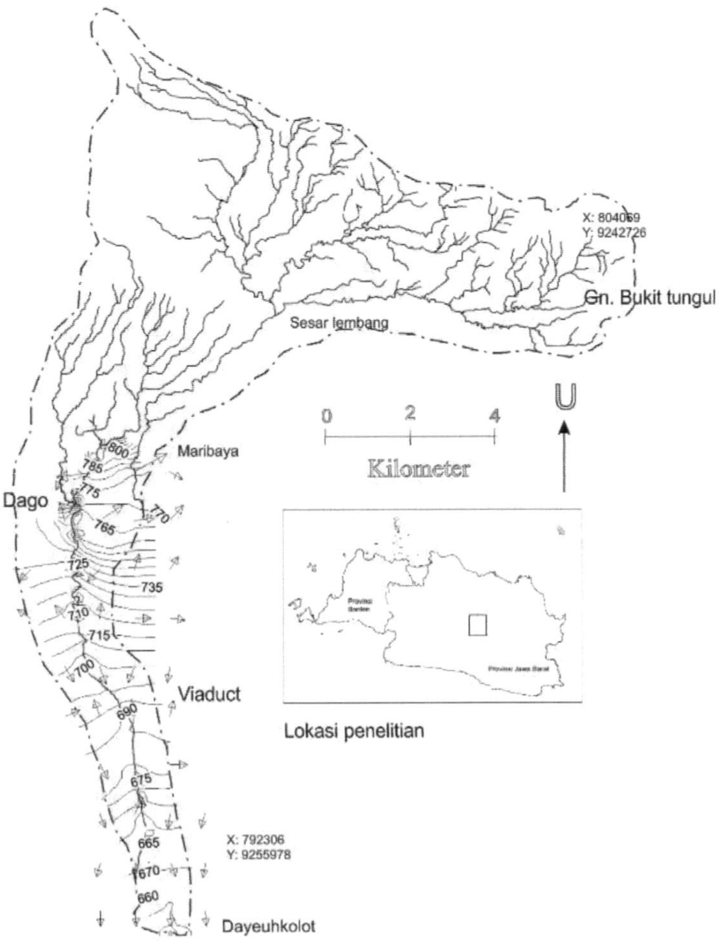

Figure 5. Results of the finite difference modelling of Cikapundung watershed.

4. Conclusions

This paper has successfully shown the relative constant state of groundwater and river interaction in the 16 years period in the Cikapundung riverbank. Therefore, the method is highly recommended to be used as a tool for water management. We will introduce this result to the authority responsible for water and environmental management. Nevertheless, small local changes are identified as one of the positive results. Both changes are due to deep wells owned by local hotel and accommodation businesses.

Acknowledgements

We wish to express our gratitude to ITB voucher scholarship for supporting this research. We also would like to thanks Prof. Dr. Lambok Hutasoit as Head of Applied Geology Group and Agus M. Ramdhan, Ph.D as the head of Hydrogeology Engineering Department for their continuous support. Lastly, we would like to thank Aditya Pratama cs., the team of undergraduate students that lend their hands and energy in the field survey.

References

[1] Darul A, Irawan DE, and Trilaksono NJ 2014 Groundwater and River Water Interaction on Cikapundung River: Revisited, Indonesia *Proceedings of the 5th International Conference on Math and Natural Sciences*

[2] Geyh 1990 Isotopic Hydrological Study in Bandung Basin, Indonesia CTA-PU Project Report

[3] Harbaugh A W, E R Banta, M C Hill, and M G McDonald 2000 MODFLOW-2000, The U.S. Geological Survey Modular Ground-Water Model - User Guide to Modularization Concepts and the Ground-Water Flow Process USGS Open-File Report 00-92. Reston, Virginia: U.S. Geological Survey, 2000

[4] Koesoemadinata dan Hartono 1981 Sedimentasi dan Stratigrafi Daerah Bandun *Prosiding PIT IAGI*

[5] Matahelumasi B and Wahyudin 2009 Penelitian Hidrogeologi Daerah Imbuhan Air Tanah dengan Metode Isotop dan Hidrokimia, di Cekungan Air Tanah Bandung Soreang, Provinsi Jawa Barat (Tahap II) (Pusat Lingkungan Hidup dan Geologi, Badan Geologi)

[6] McDonald M G and Harbaugh A W 1988 *A Modular Three-Dimensional Finite-Difference Ground-Water Flow Model: Techniques of Water-Resources Investigations of the United States Geological Survey* Book 6 Chapter A1 586 p

[7] Priowirjanto G and Marsudi 1995 Fluktuasi Muka Air Tanah di Cekungan Bandung *Prosiding Seminar Sehari Air Tanah Cekungan Bandung* Perhimpunan Ahli Airtanah Indonesia

[8] Silitonga 1973 Peta Geologi Lembar Bandung, Jawa Barat, Skala 1:100.000 Pusat Penelitian dan Pengembangan Geologi

[9] Sudjatmiko 1972 Peta Geologi Lembar Cianjur, skala 1:100.000 (Pusat Penelitian dan Pengembangan Geologi

[10] Sutrisno 1983 Peta Hidrogeologi Lembar Bandung, Jawa Barat, Skala 1:100.000 Pusat Geologi Tata Lingkungan

IOP Conf. Series: Earth and Environmental Science **29** (2016) 012027 doi:10.1088/1755-1315/29/1/012027

Groundwater Characterization of Cihaur Watershed Basin, Batujajar and Adjacent, West Bandung District, West Java, Indonesia

Fikri Noor Azy, Mohamad Sapari Dwi Hadian, and Ismawan
Geological Engineering dept., Universitas Padjadjaran,
Jalan Raya Bandung – Sumedang KM 21, 45363, Bandung, Indonesia

E-mail: fikrinoorazy@gmail.com

Abstract. The study was conducted based on data from outcrop, well data, and springs with field orientation method assisted by the use of GPS and measurement tool physical and chemical properties of groundwater. Geological conditions investigated were geomorphology and stratigraphy, geomorphology unit study area consists of four units, namely geomorphology unit of strato volcano body, foot of strato volcano, intrusion units, and plains units and the river drainage patterns are parallel and sub-parallel. Stratigraphy in the study area are volcanic breccia (Qbv), Unit Andesite (Qa), Unit Tuff (Qtf) and Unit Clay Tuffan (Qlt). The characteristics of the groundwater of the study area are in form of the physico-chemical, major elements, and hydrolic parameter of the groundwater aquifers. From 27 locations, the water quality assesment by physico-chemical properties is classified as fresh water category and based on chemical major elements, has been classified 8 facies which are located in the study area. Then, there are two lithologies which act as aquifers ie, tuff and volcanic breccias. Conductivity values in the range of volcanic breccia aquifers respectively 0,128 m/day and 0,288 m/day, transmitivity (T) ranges respectively 1,9296 m^2/day and 4,32 m^2/day. The value of conductivity in tuff aquifer is 0,063 m/day, transmitivity (T) is 0,95 m^2/day. While lithology Qlt (Clay tuffan) is lithology with very low productivity of groundwater or called groundwater rare area (akiclud) and the rock units Qa (Andesite) is a non-aquifer that is the absence of groundwater in these rock units (akifug).

1. Introduction

The need for water resource increases with the times. The use of groundwater increases due to the contamination of surface water. Therefore, the need for in-depth study on the state of subsurface water/groundwater is very important to support the management of water as part of the planning area. In order to know the geological conditions of the research area for the determination of the aquifer typology, the hydrogeological conditions of the study area and the potential of ground water in the area of research.

2. Methodology

The research area is located in Cihaur watershed basin, District Ngamprah, Batujajar, West Bandung Regency. This is an area with abundant groundwater resources, and the area is also deserves to be the area of research on the potential for groundwater around its watershed and surrounding areas. Research is done using Geological Mapping guide in order to assess the geological condition include geomorphology of study area. There 22 locations of wells and 5 location of springs are used as data manifestation of groundwater. Then, physico-chemical properties are known by measurement using a Hanna Multiparameter Probe on these manifestation of groundwater. Physical and chemical properties of groundwater is measured in the form of EC (Electrical Conductivity), TDS (Total Dissolve Solid), pH, and water and air temperature (in degrees Celsius). The method which used in determining the parameters of groundwater hydraulics, such as transmitivity and conductivity is Neuman Method Wave-Fitling [1] and [2] from drawdown pumping data with the criteria of the single well, unconfined aquifer and the calculation of T (transmitivity) is not required observation wells.

3. Result and Discussion

The results of this study are the details of geological and geomorphological data on the research location. These data will help to analyze and interprete the condition of the groundwater at the site along with the study of physical and chemical components of groundwater itself.

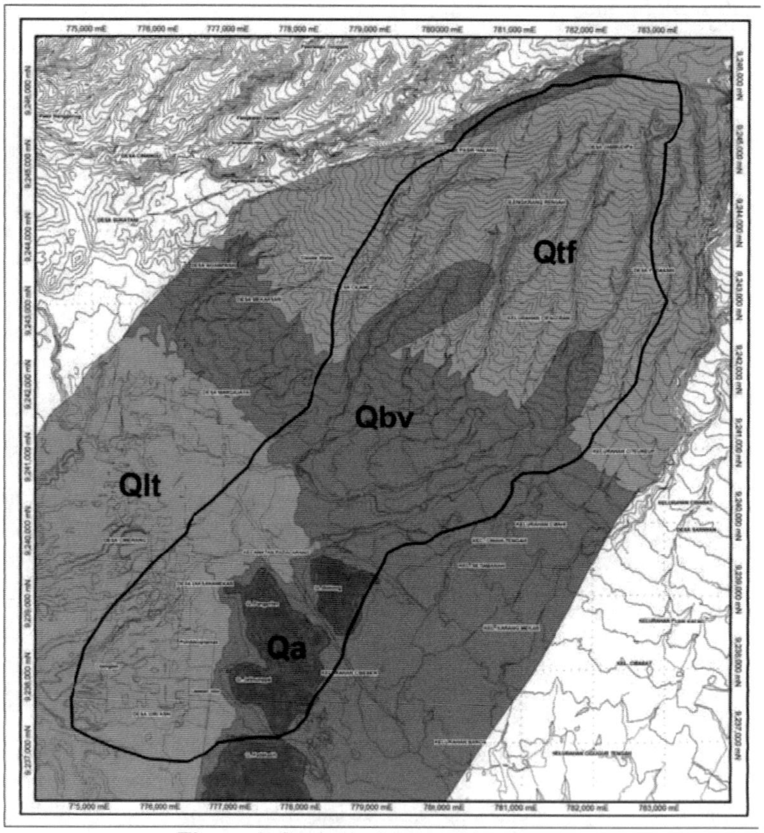

Figure 1. Geological Map of Study Area

3.1 Geomorphological and Geological Condition

The study area has a height of between 630 - 1250 m above sea level. Based on the classification slope [3] study area has a slope between 0-140%. From the data obtained, the study area is largely a volcanic morphology, consisting of mountain slopes and there are estimated lacustrine plain. Geomorphology unit division based on differences in morphography aspect, morphometry aspect, lithology properties (including facies), and geological structure in the form of activity of endogenous and exogenous, so that the unit geomorphology study area is divided into; Unit Geomorphology of Strato Volcano Body, Strato Volcano Foot, Intrusion, and Plain.

In general, the rocks in the study area consists of volcanic rocks that make up the body of volcanoes. Based on field data conducted by the Laboratory of Hydrogeology and Environmental Geology, University of Padjadjaran and compared with Regional Geological Maps of Bandung sheet by [4], in order to formed the unofficial naming of these lithologies, outside of lithostratigraphy and litodemik rocks units can be interpreted that existing lithologies research areas with the sequence of the oldest rocks to rock the youngest, namely: Unit Volcanic Breccia (Qbv), Porphyry Andesite (Qa), Unit Tuff (Qtf), and The Deposition of Lake (Qlt).

Table 1. Physico-chemical data of study area

Code	pH	TDS (mg/l)	EC (Imhos/cm)	Elevation (mdpl)	Water Temp (°)	Air Temp (°)	Type
MA-1	6.2	30	80	1195	20.8	28	Spring
MA-2	6.3	110	240	1187	23.6	26	Spring
MA-3	5.6	70	160	1174	24.7	25.5	Spring
MA-4	5.4	80	170	952	24	27.5	Spring
MA-5	6	160	220	867	25.4	28.3	Spring
SM-1	-	-	-	1175	-	-	Dug Well
SM-2	6	170	370	1074	26.2	27.1	Dug Well
SM-3	5.6	90	210	1072	23.3	27.1	Dug Well
SM-4	5	50	140	794	31.7	32.7	Dug Well
SM-5	5.4	200	420	780	25.5	25	Dug Well
SM-6	5.7	240	510	778	24.3	25	Dug Well
SM-7	6	160	340	773	24	25	Dug Well
SM-8	5.5	70	160	793	24.4	25.9	Dug Well
SM-9	4.5	180	370	843	24.8	26.6	Dug Well
SM-10	6.4	220	460	739	24.3	26	Dug Well
SM-11	6.4	230	470	747	25	31	Dug Well
SM-12	6.2	250	440	738	24	28	Dug Well
SM-13	5.8	130	270	747	25.1	27.9	Dug Well
SM-14	6.3	270	560	703	24.4	27.7	Dug Well
SM-15	5.7	440	890	699	24.4	26.7	Dug Well
SM-16	5.7	270	570	750	25.2	28	Dug Well
SM-17	5.3	60	340	853	24.7	30	Dug Well
SM-18	5.9	180	390	822	24.9	27.5	Dug Well
SM-19	5.9	220	450	817	28.5	29	Dug Well
SM-20	5.9	-	-	722	-	-	Dug Well
SM-21	5.7	650	1320	827	24.8	26.2	Dug Well
SM-22	5.8	80	190	831	24.9	26	Dug Well
SM-23	6.3	200	420	735	24.7	24.8	Dug Well

These are primary data which obtained from direct measurement from the 27 locations of the research area used a Hanna Multiparameter Probe instrument.

3.2 Groundwater Characterization

In this study, the parameters are classified is a major chemical element that is (Ca, Mg, Na, K, HCO3, SO4, Cl) and physico-chemical parameters such as pH, electrical conductivity (EC), total dissolve solids (TDS). From these data can be analyzed descriptively about groundwater characterization. On springs and wells, based on the classification [5] classified the water quality fresh water category (for TDS value 0 - 1000g / m3) and also classified as fresh water (for grades DHL 30-2000μS).

IOP Conf. Series: Earth and Environmental Science **29** (2016) 012027 doi:10.1088/1755-1315/29/1/012027

Table 2. Major chemical elements of study area

Code	Converted Laboratory Result (meq/liter)						
	Ca^{2+}	Mg^{2+}	K^{+}	Na^{+}	HCO_3^{-}	Cl^{-}	SO_4
SM-02	1.9755	0.9082	0.1352	0.2337	1.6782	1.2890	0.0922
SM-05	2.1572	0.9263	0.0833	0.5420	2.2268	0.9331	0.4095
SM-09	1.6367	1.2307	0.0615	0.3122	1.9241	1.1931	0.3419
SM-11	2.6008	0.7213	0.3770	0.2466	2.6884	0.8781	0.8780
SM-12	2.2984	0.6747	0.5463	0.1186	3.6041	0.4048	0.7726
SM-15	3.6692	2.2602	0.7946	0.3144	3.2571	3.7523	0.7947
SM-16	2.1367	1.6774	0.0891	0.8711	2.3272	2.0785	0.5143
SM-19	1.8144	1.0678	0.0720	0.8011	2.0183	1.4337	0.6263
SM-21	4.6158	4.2530	0.2514	2.2185	3.0877	4.9530	2.3402
SM-22	0.8263	0.6433	0.0581	0.2197	1.4509	0.3156	0.0421
MA-3	0.6048	0.6655	0.0691	0.1594	1.2857	0.3086	0.0581
MA-5	0.8668	0.4854	0.0534	0.1959	1.4004	0.2950	0.0473

These are primary data which obtained from the 12 locations of the research area and used bottles as samples, then analized in chemical laboratory.

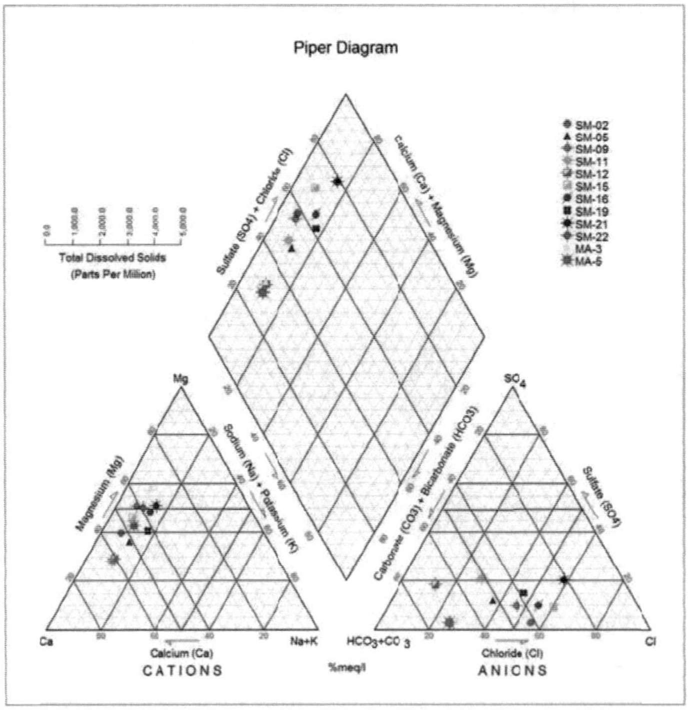

Figure 2. Modified from Piper Diagram [4] show chemical facies of study area used major elements

3.3 Major Elements

Based on the analysis using the piper diagram [4] from chemical major elements plotted, chemical facies in the study area can be grouped into 8 (eight) facies is based determination of anions and cations, namely: Facies of Calcium Chloride, Calcium Bicarbonate, Magnesium and Non dominant, Non Dominant and Chloride, Non Dominant, Magnesium Chloride, Magnesium Bicarbonate, and Non Dominant and Bicarbonate.

3.4 Hydrolics Condition of Groundwater Aquifer

Hydraulic parameters of the aquifer can be known through the pumping test. Pumping test performed on three wells owned by local residents to determine aquifer hydraulic parameters. These wells are shallow aquifer to represent basins in the study area. Hydraulic parameters of the aquifer that develops can be obtained from a pumping test in the study area were analyzed using methods Neuman Wave-Fitling use for unconfined aquifer. Wells pumping test is SM-07 represent the lithology of volcanic breccia, SM-22 represent the lithology of tuff and SM-23 represent the lithology of volcanic breccia.

The Pumping test result of SM-07 are transmitivity values 1,9296 m^2/day, conductivity values 0,128 m/day, and velocity values 0,0224 meter/day. The result for SM-22 are transmitivity values 0,95 m^2/day, conductivity values 0,063 m/day, and velocity values 0,0135 meter /day. And the result of SM-23 are transmitivity values 4,32 m^2/day, conductivity values 0,288 m/day, and velocity values 0,067 meter/day.

4. Conclusion

The characteristics of the groundwater of the study area are in form of the physico-chemical, major elements, and hydrolic parameter of the groundwater aquifers. From 27 locations, the water quality assesment by physico-chemical properties is classified as fresh water category and based on chemical major elements, has been classified 8 facies which are located in the study area.Then, there are two lithologies which act as aquifers ie, tuff and volcanic breccias. Conductivity values in the range of volcanic breccia aquifers respectively 0,128 m/day and 0,288 m/day, transmitivity (T) ranges respectively 1,9296 m2/day and 4,32 m2/day. The value of conductivity in tuff aquifer is 0,063 m/day, transmitivity (T) is 0,95 m2/day. While lithology Qlt (Clay tuffan) is lithology with very low productivity of groundwater or called groundwater rare area (akiclud) and the rock units Qa (Andesite) is a non-aquifer that is the absence of groundwater in these rock units (akifug).

Acknowledgement

The Authors would like to say thank you and highly appreciated to Hydrogeology and Environmental Geology Laboratory and Mr. Tengkuyan Iskandarsyah as project leader along with Yudhi and Aziz who have provided the data field for this paper. And we are grateful to the Water Quality Laboratory, Bandung Institute of Technology who has analyzed the major chemical parameters.

References

[1] Todd D K 1984 *Groundwater Hydrology, 2nd ed*, John Wiley & Sons, New York, USA.
[2] Kruseman GP and M A de Ridder 1994 *Analysis & Evaluation of Pumping Test Data,* Publication 47, Wegeningen, The Netherlands
[3] Van Zuidam 1985 *Aerial Photo-Interpertation in Terrain Analysis and Geomorphology Mapping*. Publisher The Hague, Netherland.
[4] Piper A M 1944 *A Graphic Procedure in The Geochemical Interpretation of Water Analysis.* Trans, Am. Geophys. Union, Washington, D.C.
[5] Freeze R A and Cherry J A 1979 *Groundwater*. Prentice-Hall, Inc. Englewood Cliffs, New Jersey, USA.

Rock surface roughness measurement using CSI technique and analysis of surface characterization by qualitative and quantitative results

Husneni Mukhtar[1], Paul Montgomery[1], Gianto[1] and K. Susanto[2]

[1]ICube Laboratory, University of Strasbourg-CNRS, 23 rue du Loess, 67037 Strasbourg, France
[2]Geophysics, Universitas Padjadjaran, Jl. Raya Jatinangor-Sumedang km.21, Indonesia

mhusneni@unistra.fr

Abstract. In order to develop image processing that is widely used in geo-processing and analysis, we introduce an alternative technique for the characterization of rock samples. The technique that we have used for characterizing inhomogeneous surfaces is based on Coherence Scanning Interferometry (CSI). An optical probe is first used to scan over the depth of the surface roughness of the sample. Then, to analyse the measured fringe data, we use the Five Sample Adaptive method to obtain quantitative results of the surface shape. To analyse the surface roughness parameters, H_{mm} and R_q, a new window resizing analysis technique is employed. The results of the morphology and surface roughness analysis show micron and nano-scale information which is characteristic of each rock type and its history. These could be used for mineral identification and studies in rock movement on different surfaces. Image processing is thus used to define the physical parameters of the rock surface.

1. Introduction

The surface analysis of rocks is necessary in order to be able to define their physical parameters, such as for instance the surface roughness that can be measured by microscopy to give information not only about the rock surface texture but also on its porosity. Various techniques that have been used to obtain images of rock surfaces are conventional optical microscopy, electron microscopy such as scanning electron microscopy (SEM), atomic force microscopy (AFM), etc. The results of image measurements using conventional optical microscopy is on the micro-metre scale and for SEM it is up to the nano-scale.

CSI, or white light scanning interferometry (WSLI), or Coherence Probe Microscopy (CPM) as it also known, is an increasingly popular method for measuring the roughness and shape of microscopic surfaces. The technique consists of a fast, non-contacting method for measuring surfaces that is increasingly important as a means of examining not only material surfaces but also components such as miniature optical elements, micro-fluidic devices and micro-electro-mechanical systems. It combines the axial resolution of an interferometer and the lateral resolution of a high-magnification optical microscope; commercial instruments are now an integral part of most research laboratories [1-4].

Most real surfaces are smooth surfaces which have higher self-affinity. They can be treated as fractals under well-defined spatial limits, so that within these limits specific statistical properties upon scale change are preserved by the surface morphology. In this case, they can be auto-scaled [5,6]. Fractals can be characterized by self-affinity, an example of these in nature being trees, leaves, clouds, and rocks. Self-affinity occurs when the fractals are not exact copies of themselves but there is still preservation of the general shape with more than one transformation and the need for a scale reduction for morphology invariance [5,7]. According to Peitgen et al. [5,8], one of the most common methods that is used to calculate the fractal dimension is the box counting method or what we call in this paper, the window resizing technique.

2. Coherence Scanning Interferometry (CSI) Technique

CSI is an important optical technique that is now widely used in the measurement of surface roughness and microscopic surface shape. CSI has the advantages of being rapid, non-destructive and applicable to many different types of surfaces. The technique makes use of a series of white light fringes superimposed on an image of the sample on the camera target, the central fringe along the optical axis corresponding to the position of the surface. The fringes are scanned over the whole depth of the sample surface by modifying the distance between the objective and the sample. A series of images is acquired with a camera at regular intervals to give a stack of XYZ images and using signal processing along the z-axis the peak of the fringe envelope is determined at each pixel and thus the corresponding height of the surface at each point in the image [9].

The system developed at ICube is shown in figure 1, based on a modified Leitz-Linnik microscope equipped with two 50× objectives and an incandescent lamp (centered at a wavelength of 580 nm), giving a lateral optical resolution of 0.42 μm with a CCD camera Basler-AVA 1000, and using a piezo scanner (PIFOC from PI) for vertical scanning. The computer specification is Processor Intel® Xeon® CPU 2.33 GHz RAM 8G.

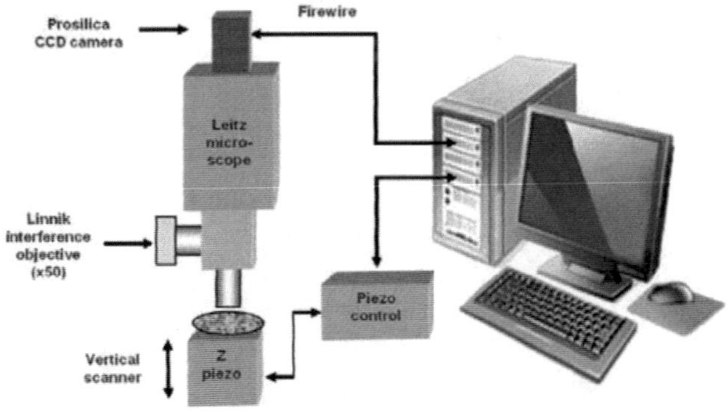

Figure 1. CSI instrument developed at ICube [10]

3. Analysis Methods

3.1. Five sample adaptive (FSA)

The FSA method (fringe visibility method proposed by Larkin [11]) is used to determine the fringe envelope and the surface height. This method detects the envelope of fringe signals by using five sampling positions along the optical axis. For each sampling position on the envelope, the amplitude value is calculated from the local position and four neighbouring sampling positions. The calculated envelope is given by:

$$4\,A^2\,sin^4\,\varphi = (I_2 - I_4)^2 - (I_1 - I_3)(I_3 - I_5) \quad (1)$$

where A is the amplitude value on one sampling position; φ is phase shift due to scanning step; I_3 is the local position; I_1, I_2, I_4, I_5 are the neighboring sampling positions.
For values of φ close to 90°, the equation (1) can be written as:

$$A^2 = \frac{1}{4}[(I_2 - I_4)^2(I_1 - I_3)(I_3 - I_5)] \quad (2)$$

The amplitude value of the envelope A can be calculated by only two multiplications and one square root operation. This gives the advantage in time computation. However, the FSA algorithm requires the values of φ to be close to 90°, for accurate measurement.

3.2. Window resizing technique

In order to analyse the roughness data quantitatively, we used the window resizing technique to give the roughness amplitude parameters, H_{mm} or the peak-valley roughness and R_q or the root mean square (RMS) value of the roughness as a function of window size. The roughness calculation is then conducted on the cell. The number of squares depend on the value of δ, where $\delta = 1, 2, 3, \cdots, mod(n/2)$ where n is a pixel size in 1D. This is called a geometrical series and is written as:

$$\text{window size} = 2^{\delta+1}\ \text{squares} \quad (3)$$

The cell has the shape of a square, where $\partial x = \partial y$. Window resizing calculates the roughness values at a given cell size over the whole of the image arithmetically.

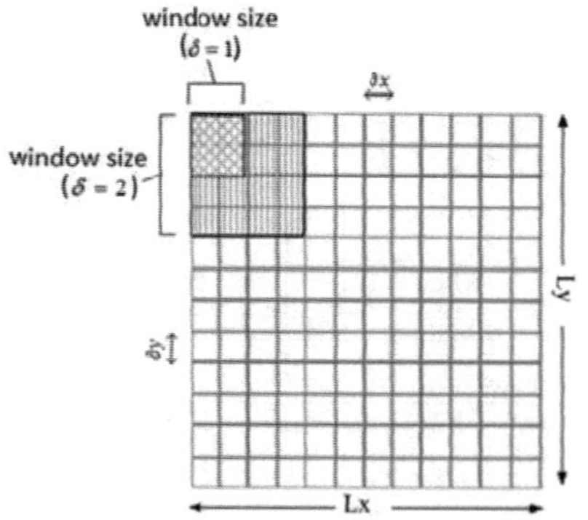

Figure 2. Composition of one cell consisting of four squares for $\boldsymbol{\delta = 1}$

We use the formula in equation (4) to calculate H_{mm} and equation (6) to calculate R_q [12]

$$H_{mm}(\delta) = \frac{1}{(i_{max}-2\delta)(j_{max}-2\delta)} \sum_{i=1+\delta}^{i_{max}-\delta} \sum_{j=1+\delta}^{j_{max}-\delta} (z_{max} - z_{min})_{i,j} \quad (4)$$

where

$$(z_{max})i,j = Max(z_{n,m}), n \in [i - \delta, i + \delta], m \in [j - \delta, j + \delta]$$
$$(z_{min})i,j = Min(z_{n,m}), n \in [i - \delta, i + \delta], m \in [j - \delta, j + \delta] \quad (5)$$

$$R_q(\delta) = H_{rms}(\delta) = \frac{1}{(i_{max}-2\delta)(j_{max}-2\delta)} \sum_{i=1+\delta}^{i_{max}-\delta} \sum_{j=1+\delta}^{j_{max}-\delta} \sqrt{\langle (z - \bar{z})^2 \rangle_{i,j}} \quad (6)$$

and

$$\langle (z - \bar{z})^2 \rangle_{i,j} = \frac{1}{(1+2\delta)^2} \sum_{n=i-\delta} \sum_{m=j-\delta} \left(z_{n,m} - \bar{z}\right)^2 \quad (7)$$

With:

$$(\bar{z})_{i,j} = \langle z \rangle_{i,j} = \frac{1}{(1+2\delta)^2} \sum_{n=i-\delta}^{1+\delta} \sum_{m=j-\delta}^{j+\delta} z_{n,m} \quad (8)$$

4. Results : Surface Characterization

Two types of limestone, marble limestone and hematite limestone, were used as samples in order to test the CSI method. These stones have very fine and smooth surface because of their self-affinity character. A coating treatment process was not necessary for each sample as in the case of SEM measurement, which means that using the CSI method, the samples are not damaged. Results of CSI roughness measurement are presented in size dimension of 1024×1024 [pixel size dx = 0.13 μm].

4.1. Results of sample marble limestone

Figure 3 shows a gray scale height image in (a) and a 3D altitude distribution in (b). A line profile A-B from (a) is shown in (c).

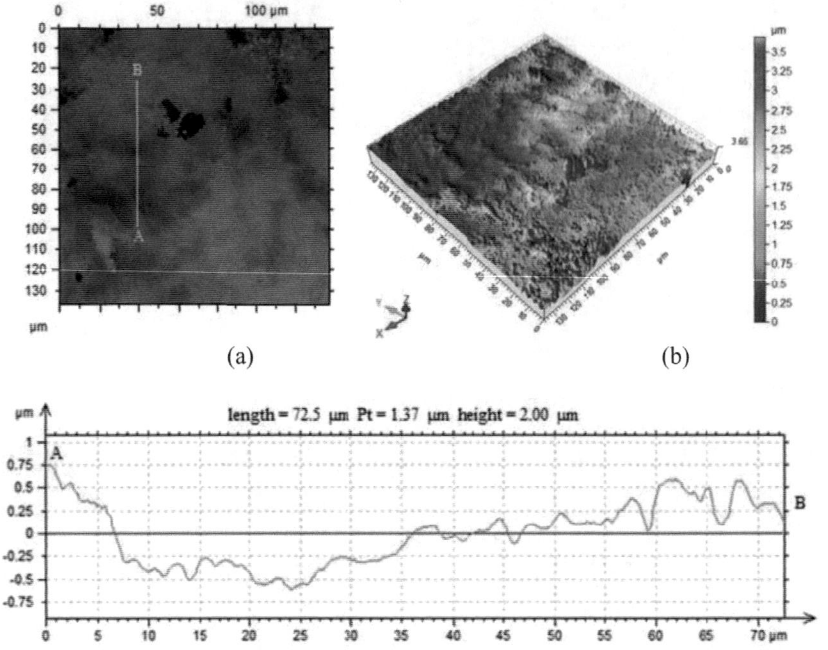

(a) (b) (c)

Figure 3. Marble limestone roughness measured by CSI: (a) gray level height image (b) 3D altitude distribution (c) line profile along line A to B from (a)

The calculation result of the roughness distribution according to window size is shown in figure 4 based on the image in figure 3(c).

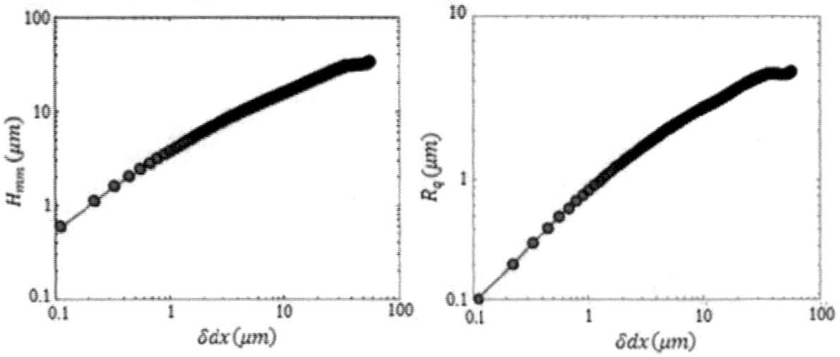

(a) (b)

Figure 4. Result of window resizing analysis for marble limestone surface rock (a) peak-valley roughness (b) RMS roughness

4.2. Result of sample hematite limestone

Figure 5 shows a gray scale height image in (a) and a 3D altitude distribution in (b). A line profile A-B from (a) is shown in (c).

IOP Conf. Series: Earth and Environmental Science **29** (2016) 012028 doi:10.1088/1755-1315/29/1/012028

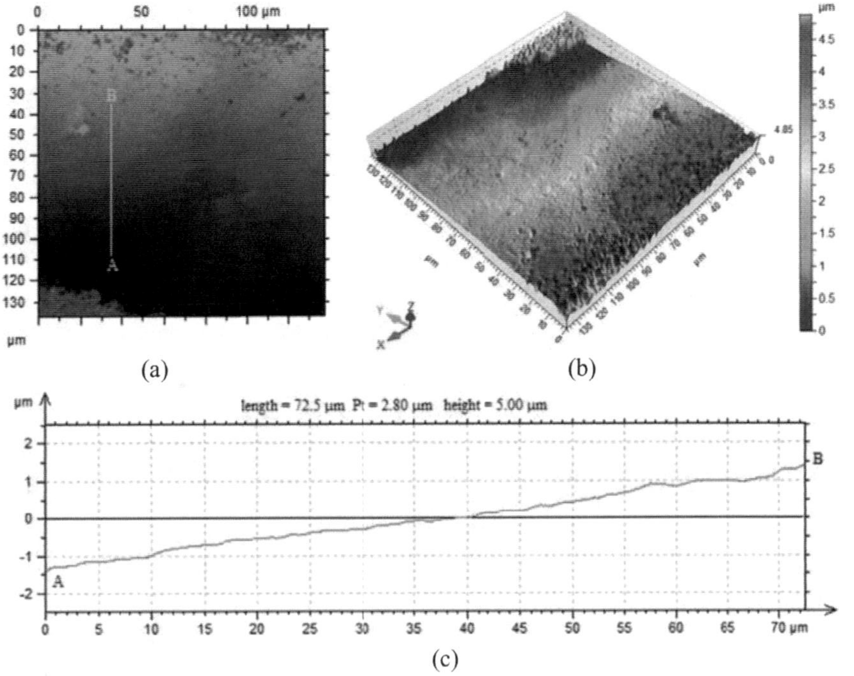

Figure 5. Hematite limestone measured by CSI (a) gray level image (b) altitude distribution (c) line profile A-B from (a)

This image shows fine texture, and is smoother than the sample in figure 3. The line profile presented in figure 5 shows little microstructure. The calculation result of the roughness distribution according to window size is shown in figure 6 based on the image in figure 5(a).

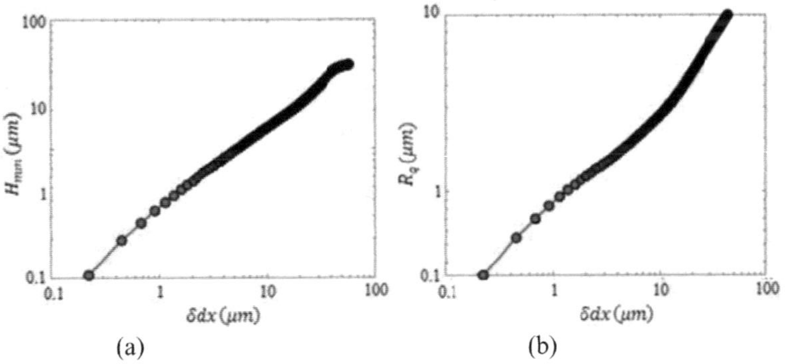

Figure 6. Result of window resizing analysis for hematite limestone surface rock (a) peak-valley roughness (b) RMS roughness

Conclusion

The CSI technique can be used to quantitatively measure the surface roughness of rock sample surfaces. Image processing can thus be used to define the physical parameters of the rock surface. In future work, the analysis of the results of the images from the CSI technique could be used for mineral identification and to study the link between surface roughness and the movement of rocks on different types of surfaces in the study of landslides.

References

[1] Lee B S and Strand T C 1990 Profilometry with a coherence scanning microscope. Appl *Opt.* **Vol.29,** 3784–3788.

[2] De Groot P and Deck L 1995 Surface profiling by analysis of white-light interferograms in the spatial frequency domain. *J. Modern Opt.* **Vol.42,** 389–401.

[3] Sheppard C J R, Roy M and Sharma M D 2004 Image formation in low-coherence and confocal interference microscopes. *Appl. Opt.* **Vol.43**, 1493–1502.

[4] J M Coupland and J Lobera 2009 Measurement of steep surfaces using white light interferometry *J Strain*, **Vol.46**, 69-78.

[5] M Prado, L C Lima and R A Simão 2012 Scale laws for AFM image evaluation: potentialities and applications. Current Microscopy Contributions to Advances in Science and Technology, *Microscopy Series No.5* **Vol.2** (A. Méndez-Vilas, Ed.), Spain: Formatex Research Center C/Zurbarán 1,2°-Officina 1 06002 Badajoz.

[6] Mandelbrot BB 1983 *The fractal geometry of nature. 1st ed.* San Francisco, Freeman and Company.

[7] Mandelbrot BB 2001 *Gaussian Self-Affinity and Fractals. 1st ed.* New York: Springer.

[8] Peitgen H et al 1992 *Fractals for the classroom, part 1: Introduction to fractals and chaos. 1st* ed., New York: Springer.

[9] Montgomery P, Montaner D, Salzenstein F, Serio B. & Pfeiffer P 2013 Implementation of a fringe visibility based algorithm in coherence scanning interferometry for surface roughness measurement. *SPIE*, **Vol. 8788**, 124-126.

[10] E Pecheva, P Montgomery, D Montaner, L Pramatarova 2007, White light scanning interferometry adapted for large area optical analysis of thick and rough hydroxyapatite layers, *Langmuir* **Vol.23**, 3912–3918.

[11] Kieran G Larkin 1996 Efficient nonlinear algorithm for envelope detection in white light interferometry. *J. Opt. Soc. Am.* A/ **Vol. 13**, No. 4, 832-843.

[12] KL Apedo et.al 2015 Cement paste surface roughness analysis using coherence scanning interferometry and confocal microscopy. *Materials Characterization* **Vol.100**, 108–119.

IOP Conf. Series: Earth and Environmental Science **29** (2016) 012029 doi:10.1088/1755-1315/29/1/012029

Controlling Groudwater System by Pattern Fracture Approach in Subsurface Volcanic Deposit: Mt.Salak-Mt.Pangranggo, West Java, Indonesia

Cipta Endyana[1], Hendarmawan[2], Emmy Sukiyah[3], Irwan Ary Dharmawan[4]

[1,2,3] Faculty of Geology, Universitas Padjadjaran, Indonesia
[4] Departement of Geophysics, Faculty of Mathematics and Natural Sciences, Universitas Padjadjaran, Indonesia

E-mail: cipta.endyana@unpad.ac.id

Abstract. In general, the volcanic region has high potential of water resources. However, volcanic field are very complex in structure and texture of rock compared with sedimentary rocks. They also have different porosity in type and distribution, with the spread of the highly different within a short distance. Consequently, groundwater in this volcanic area is quite difficult to predict groundwater flow. Those rocks should be identified as vertical and lateral spreading. In fact, the groundwater a is not flowing in the volcanic rock pores only, but also flow in fractures that developed by the volcanic and tectonic processes. Ciherang area which is located between Mount Salak and Mount Gede-Pangrango and bypassed by tectonic faults, has a complex fracture pattern. The result of this fracture pattern research indicated that at least four pattern of fracture systems were developed. All fracture patterns were suggested in relation with the imposition of volcanic rocks. Groundwater in these fractures have to consider for water resources calculation. Therefore fracture media has become one of the important parameters in the calculation of water resources. The modelling of subsurface volcanic deposit was developed by resistivity value of rock deposits. They can describe the distribution of volcanic deposits until 150 meters below surface. The fracture that constantly developed up to certain depth will be exposed by contrast enhancement of resistivity model of rock deposits. Delineation of fracture pattern will be known which each fracture pattern is associated with the flow of groundwater. Furthermore, there are also fractures are influenced by tectonic faults, and fractures caused by both of the processes. Fractures with high intensity indicated direction of porous media have a trend with relatively north-south direction and the fracture which is constantly up to certain depth is indicated as a pathway of groundwater flow, but several fractures which are affected by tectonic regional process will become a barrier of groundwater flow. To control the validity of fracture development, the stable isotope deuterium and oxygen content were used in groundwater as well as flowing water between one fracture systems to other systems.

1. Introduction

Mount Salak is a mountain that formed by volcanism process in quaternary period. This volcanic area has a high potential of groundwater. Mount Salak has a large content of groundwater caused by high fracture intensity. The fractures in volcanic areas have contributed for great potential groundwater resources

Fracture patterns mapping is important to gain knowledge of the fracture system prevailing in this area, both of which are influenced by tectonic fractures or pattern formed on the clotting time of volcanic rocks. As revealed by Denny et.al in his paper on methods of structure mapping that control the aquifer, the fracture as a structural element of regional or local geology strongly influence the determination and the presence of groundwater recharge zone [1].

Some results provide two-dimensional form of lineaments in the valleys and hills of analog or digital maps, satellite imagery and DEM, with lineament analysis used as reference to determine the cracks intensity level of an area [2]. Research area was shown at Figure 1.

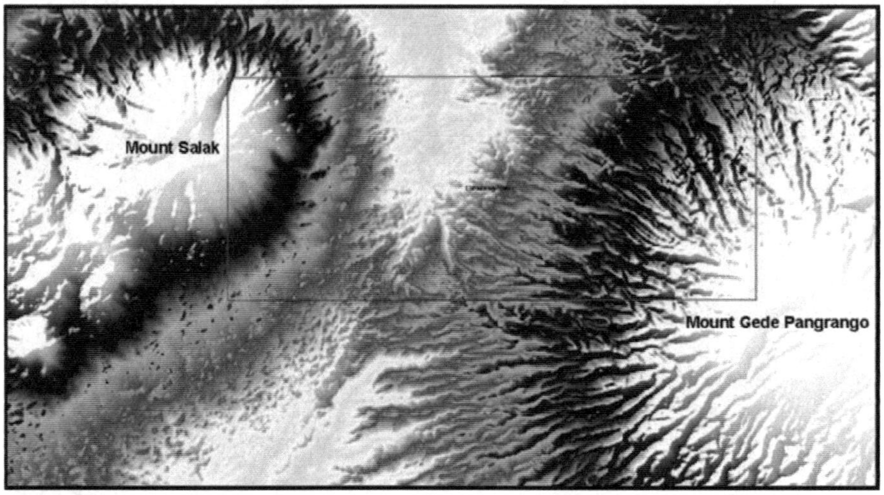

Figure 1. Research Area located between Mount Salak and Mount Gede Pangrango at Ciherang (indicated by red box), West Java, Indonesia

2. Regional Geologic Setting

Ciherang areas included in Bogor Sheet on regional geological maps. Based on data on regional geological map is known the oldest rocks exposed in the study area consists of tufa pumiceous which belong to the old volcanic rock units that spread in the east, north and south areas of research. Other lithology is also in the form of andesitic basalt lava, which belongs to the volcanic deposits of Pangrango Unit. In the west of the study area is composed of lava deposit, tuffaceous breccias and lapilli, basaltic andesite lava flows, which belong to th volcanic deposits of Mount Salak Unit. These volcanic deposit units belong to the old Quaternary volcanic deposit. (Figure 2).

Tectonic activity in this area began in the Early Tertiary, followed by Plio-Pleistocene tectonic activity that reactivate the product tectonic Early Tertiary period. These events resulted in the establishment of faults which have generally northeast - southwest and northwest - southeast direction. Fractures that has been formed become a weak zone for young volcanic quaternary rock.

Base on regional geological study, research area are generally composed by rock quarter groups, in which the existing deposits deposition is a young volcano that can not be separated between sandy pumice tuff and tuffaceous lava breccia derived from deposition Pangrango. This young volcanic sediment has a thickness sufficient. Then, under the quarter rock group, that old volcanic rocks group has been deposited composed of andesitic basaltic breccia, andesitic lava, tuffs and agglomerates which is the inseparable deposition. In the southern part of research area, there is a fine-coarse clastic sediments as a tertiary rocks which have been folded and faulted by tectonics [4].

Most researchers agreed that the volcanic eruption in this area relates to the activity of plate collision [5]. Based on geophysical data, it is known that the subduction pathways in the southern island of Java, is a continuous zone, but the volcanic eruption on the surface is separated from one another. This phenomenon indicates that the activity magmatism or volcanism does not by itself come to the surface, but there must be some fractures that have a function as a media to release the magma to the surface. Fracture is always a fault lines, for example on the island of Java, the position of the active volcanoes are in the volcanic arc environment on the surface appearance was in the Baribis Fault Zone [5].

With geological facts described above, it is in Java which is also found volcanic (from the age of Paleogene to the present) associated with fault zones confirmed [6]. Examples Katili and Sudradjat connect the presence of volcanoes around West Java and the surrounding areas with Cimandiri Fault zones and fault Citanduy. Some other examples are the appearance of Volcano Ciremai in Cirebon-Kuningan which is at the intersection of the fault zone and Fault Baribis Citanduy; Volcano Tangkubanprahu-Burangrang are in fault zones Cimandiri; and Mount Krakatau in the Sunda Strait was in the Sunda Strait fault zone [7].

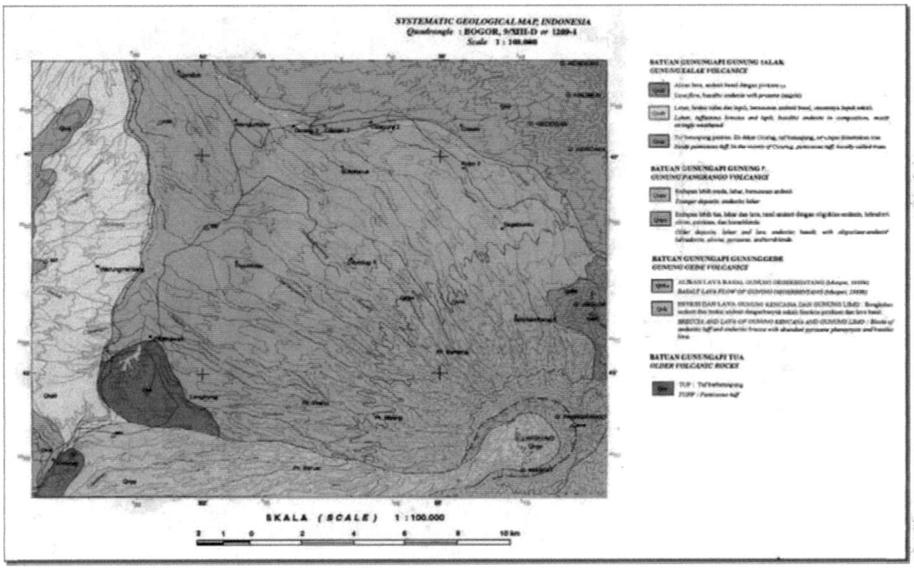

Figure 2. Regional Geological Map of research area-Bogor sheet [3].

3. Method

Fractures can define easily with satellite image such as Landsat, Radarsat, or Digital Elevation Model (DEM). There are a lot of fracture tracks provide by lineament in surface, in this case the lineament of hills and valley. The surface method to calculate of fractures intensity and delineate of fracture pattern is used Frequency Domain Decomposition (FDD) [7]. FDD is useful for lineament analysis. The result of FDD is the intensity of distribution of fractures in research area, not only the value of intensity but also pattern of fracture distribution and the main direction of tectonic or volcanic styles. Those fractures were appear in DEM has a pattern and can be described the force that occurred in research area. The fractures patterns can be classified according to the style influence that occurs in the research area, whether the fractures pattern caused by effect of tectonic, volcanic, or both processes [7, 8].

Geo-electric method used to determine the condition of the distribution of subsurface lithology. Resistivity value produced by the geo-electric instruments very well and efficiently used to determine the condition of groundwater and subsurface rocks in volcanic deposits. Its value is highly dependent on several factors, namely groundwater, salinity, saturation, and lithology aquifer. Resistivity method is always used to solve many problems such as groundwater to determine the depth, thickness and boundaries of the aquifer. Rock resistivity value reflects the hardness of rocks in accordance with

Table-1 in relative terms, with known resistance value and classified rock type. It can be predicted subsurface rock conditions up to a depth of 150 meters. [10]

Table 1. Resistivity value for each lithology at volcanic deposit at research area.

Material Value	Resistivity Range	Typical
Igneous &Metamorphic	10^2-10^8	10^4
Sedimentary Rock	10-10^8	10^3
Unconsolidated	10-1-10^8	10^3
Groundwater	10^{-1}-10^4	5
Pure water	1-10	10^3

Resistivity (Ωm)		Lithology Interpretation	Hydrogeology Interpretation
6	19	Welded tuff	Impermeable Layer
20	30	Fine tuff	Impermeable Layer
31	40	Medium tuff	Impermeable Layer
41	60	Coarse tuff	Aquifer
61	90	Lapili Tuff	Aquifer
91	200	Gravelly sand/Lapilli Tuff	Aquifer
201	400	Breccia matrix supported	Aquifer
401	900	Breccia grain supported	Impermeable Layer

4. Fracture Pattern of Mt. Salak-Mt. Pangrango

As a detail result of surface fracture pattern analyzing, there are several fractures pattern have been developed by volcanism and tectonic processed in this research area. There are at least four different types of fracture pattern in Mount Salak area, which are formed by volcanic and tectonic processed. First type formed by volcanic processed, second and third types formed by both of volcanic and tectonic processed, and four type is affected by tectonic processed. Figure 3

The four types of fracture patterns described as follows:

- Type A, is affected by volcanic process, has a relatively North-South direction, good porosity of aquifer.
- Type B, is affected by both volcanic and tectonic process, has a relatively Northwest-Southeast direction, good porosity, but in a few place are impermeable.
- Type C, is affected by both volcanic and tectonic process, has a relatively East-West direction., good porosity, but in a few place are impermeable
- Type D, is affected by tectonic process, has a relatively the same direction with trend of Pelabuhan Ratu fault along research area, and the possibility of this fracture pattern is impermeable in many locations.

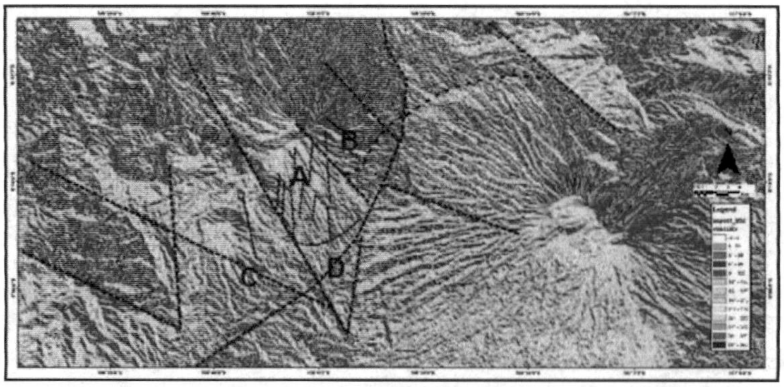

Figure 3. Four types of Surface fracture pattern in Ciherang research area

By using resistivity model as vertically analysis. We obtained an overview of distribution of subsurface volcanic deposits of the study area. In the Figure 4 below describe that fracture pattern occurrences with northwest-southeast direction, represented by the red line, the fracture pattern that have characterized as a media of fractures to be a pathway of groundwater flow. In addition to the fracture pattern, there is a fracture pattern with relatively the north-south direction has character of fracture as a barrier of groundwater flow. The fracture not only appears on the surface, but constantly appears until at depths greater than 50 meters Figure 5 [12].

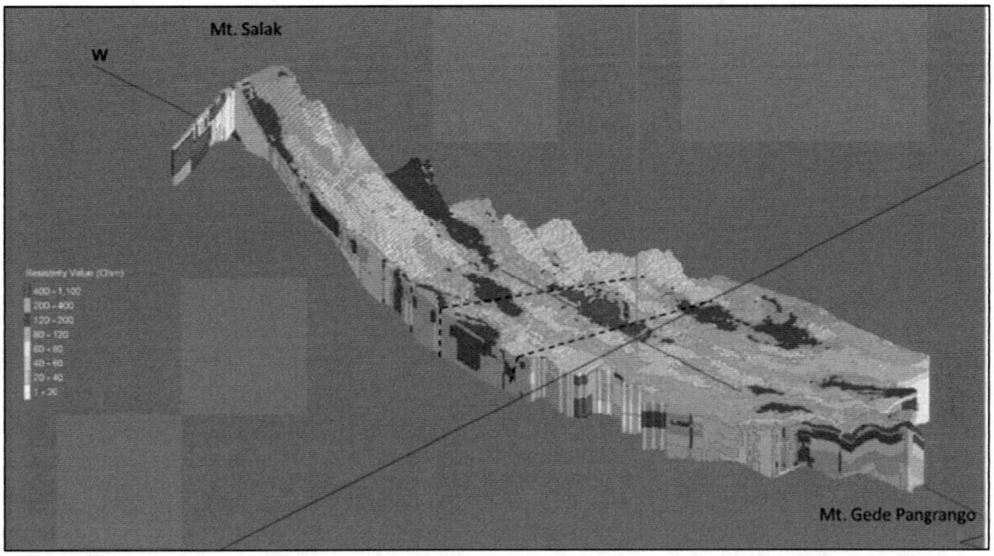

Figure 4. Model of Iso-resistivity describe the characteristic of fracture pattern in subsurface.

5. Results and Discussions

Volcanism and tectonic process has highly affected in research area. The influence of their processed made many different in developing of each fracture pattern characterization. The fracture media become an important parameter for groundwater, which is fracture media as an secondary porosity in volcanic aquifer system. In the early stages of research, lineament analysis can be used to determine the fracture pattern on the surface morphology of the research area. Based on this research the fracture pattern with affected by volcanic processed have a better porosity then fracture pattern developed with affected by regional fault.

Actually, there are many fractures that develop in research area as a laterally or vertically, but in this research describe only for extreme fracture that really influence to the flow of groundwater. As an early stage of research related to the fracture pattern controlled groundwater flow in volcanic system at surface area, on the stage of the research of fracture patterns that occurs in the subsurface area. The purpose of this research is to determine the fracture pattern vertically in subsurface area. Determination of fracture that continues until subsurface becomes more essential to know the groundwater flow system in volcanic rocks.

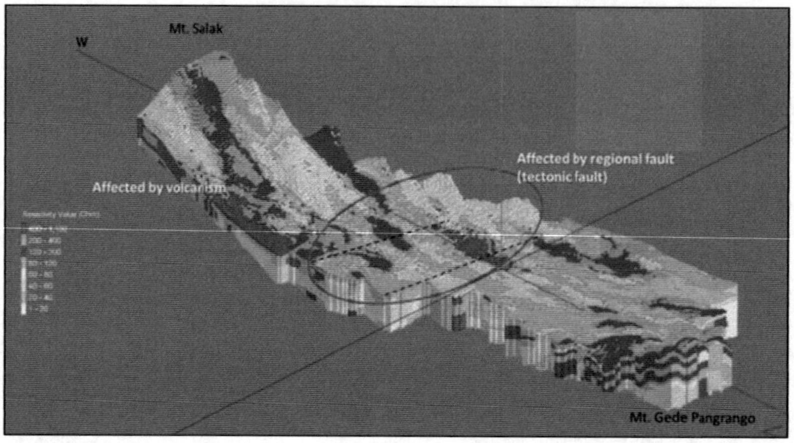

Figure 5. Two different characteristics of fracture pattern with different effect. The tectonic involve for fracture pattern indicated by red circle as a barrier of groundwater flow.

References

[1] Denny SC, Allen DM and Journey JM 2007 DRASTIC-Tm: a modified vulnerability mapping method for structurally controlled aquifers In the southern Gulf Islands, British Columbia, Canada. *Hydrogeology Journal* **15**:483-493. (Springer-Verlag)

[2] Sander P 2007. Lineament in groundwater exploration: a Review of applications and limitations. *Hydrogeology Journal* **15**:71-74. (Springer-Verlag)

[3] Efendi AC and Hermanto B 1998 *Peta Geologi Regional Lembar Bogor*, skala 1:100.000. (Pusat Penelitian dan Pengembangan Geologi, Bandung)

[4] Muertianto E 2006 *Peta Hidrogeologi Regional Lembar Bogor*, skala 1:100.000. (Pusat Penelitian dan Pengembangan Geologi, Bandung)

[5] Haryanto I 2013 Struktur Sesar di Pulau Jawa Bagian Barat Berdasarkan Hasil Interpretasi Geologi. *Bulletin of Scientific Contribution* (Geologi Unpad) **11** 1

[6] Hamilton W 1979 Tectonics of the Indonesian Region: *Geological Survey Professional Paper 1078*, (US. Government Printing Office)

[7] Nyborg M, Berglund J and Triumf CA 2007 Detection of lineament using airborne laser scanning technology: Laxemar-Simevarp, Sweden. *Hydrogeology Journal* **15**:29-32 (Springer-Verlag)

[8] Singhal BBS and Gupta RP 2010 *Applied Hydrogeology of Fractured Rocks*. (Springer London)

[9] Endyana C, Hirnawan F, Hendarmawan and Mardiana U 2011 Pendugaan nilai tahanan jenis batuan sebagai upaya untuk mengetahui struktur geologi yang berkembang pada endapan vulkanik di Kec. Padarincang, Provinsi Banten. *Buletin Sumber Daya Geologi* **6** 3

IOP Conf. Series: Earth and Environmental Science **29** (2016) 012030 doi:10.1088/1755-1315/29/1/012030

Numerical Simulation of non-Newtonian Fluid Flows through Fracture Network

I A Dharmawan[1], **R Z Ulhag**[1], **C Endyana**[2] **and M Aufaristama**[1,3]

[1] Department of Geophysics, Universitas Padjadjaran, Indonesia
[2] Faculty of Geology, Universitas Padjadjaran, Indonesia
[3] Institute of Earth Sciences, University of Iceland, Sturlugata 7, Reykjavik 101, Iceland

E-mail: iad@geophys.unpad.ac.id

Abstract. We present a numerical simulation of non-Newtonian fluid flow in a two-dimensional fracture network. The fracture is having constant mean aperture and bounded with Hurst exponent surfaces. The non-Newtonian rheology behaviour of the fluid is described using the Power-Law model. The lattice Boltzmann method is employed to calculate the solutions for non-Newtonian flow in finite Reynolds number. We use a constant force to drive the fluid within the fracture, while the bounceback rules and periodic boundary conditions are applied for the fluid-solid interaction and inflow outlflow boundary conditions, respectively. The validation study of the simulation is done via parallel plate flow simulation and the results demonstrated good agreement with the analytical solution. In addition, the fluid flow properties within the fracture network follow the relationships of power law fluid while the errors are becoming larger if the fluid more shear thinning.

1. Introduction

The study of fluid flow through fracture system is encountered in many industrial problems, such as solute transport problem [1, 2], enhanced oil recovery (EOR) [3] and many more. Solute transport in rock fractures are great attention for environmental problems, for instance remediation of contaminated of groundwater flow. In EOR processes, the non-Newtonian fluids like polymer and surfactant are play important roles to enhance the oil productivity. In those cases, the analytical solutions do not exist due to the complexity of the flow, hence the numerical simulation is promising tool to estimate the physical behavior of the system.

We employed a lattice Boltzmann method (LBM) for the simulation of non-Newtonian fluids. Recently, LBM has long standing good achievements for solving Navier-Stokes equations including heat transfer problems [4]. Compared with other traditional method like finite difference and finite element method, LBM is based on kinetic theory approach, hence it has some advantages especially when we deal with some complicated transport system in complex geometries [5]. On the other hand, LBM algorithm is also easy for parallel computation [6].

In this paper, a non-Newtonian fluid based on LBM model will be implemented in a fracture network. The power law rheology model is used to describe the non-Newtonian properties of the fluid [8, 9]. The self-affine fracture network was generated by performing fractional Brownian Motion (fBM) method which proposed by Madadi et al [7]. Hence, it can capture the surface roughness of the natural fracture by including the fractal dimension. While, the LBM simulations of Newtonian flow for this system has been clearly discussed in [10, 11, 12]. The objective of the present work is to analyse numerically of the flow velocity, shear stress and viscosity and its dependence on the fluid rheology and the fracture geometry effect.

2. The lattice Boltzmann method

The LBM method was developed as an alternative tool for Computational Fluid Dynamics. LBM consists of discrete particles with particular velocities. The particle distribution functions is defined as $f_i(\vec{x}, t)$, where $\vec{x}$ and t is position vector and time, respectively. The subscript i is the lattice link of the distribution function. During simulation, the particles will propagate and collide in a regular lattice. The distribution functions will evolve and interact in such way that that mass, momentum and energy are conserved.

The evolution equation of the LBM is defined as [13]

$$f_i(\vec{x} + \vec{c_i}\Delta x, t + \Delta t) = f_i(\vec{x}, t) + \Omega_i(\vec{x}, t) \tag{1}$$

Here, in this paper we consider D2Q9 lattice, where D2 indicates for two dimension and Q9 indicates for nine velocity vectors. The discrete velocities for this lattice are the following

$$\begin{aligned} \vec{c_i} &= (0,0); \quad i = 0 \\ \vec{c_i} &= \cos(\frac{\pi}{2}(i-1)), \sin(\frac{\pi}{2}(i-1)); \quad i = 1,2,3,4 \\ \vec{c_i} &= \sqrt{2}\cos(\frac{\pi}{2}(i-1) + \frac{\pi}{4}), \sin(\frac{\pi}{2}(i-1) + \frac{\pi}{4}); \quad i = 5,6,7,8 \end{aligned} \tag{2}$$

The Ω_i is the collision operator. For this simulation, we use Bhatnagar-Gross-Krook collision operator [14]

$$\Omega_i = \frac{-1}{\tau}[f_i(\vec{x}, t) - f_i^{eq}(\vec{x}, t)] \tag{3}$$

while τ is the time of relaxation. The equilibrium distribution function f_i^{eq} is given by

$$f_i^{eq}(\vec{x}, t) = \omega_i \rho \left(1 + 3\vec{c_i} \cdot \vec{u} + \frac{9}{2}(\vec{c_i} \cdot \vec{u})^2 - \frac{3}{2}\vec{u}^2\right) \tag{4}$$

where $\omega_0 = \frac{4}{9}$ for $i = 0$, $\omega_1 = \frac{1}{9}$, for $i = 1,2,3,4$ and $\omega_i = \frac{1}{36}$ for $i = 5,6,7,8$. The macroscopic fluid parameters like fluid density ρ and velocity $\vec{u}$ can be calculated from the distribution functions at each node by

$$\rho = \sum_i f_i \quad \text{and} \quad \rho\vec{u} = \sum_i f_i \vec{c_i} \tag{5}$$

The kinematic viscosity is related with relaxation time, which defined as

$$\nu = \frac{1}{3}\left(\tau - \frac{1}{2}\right) \tag{6}$$

The relation between stress tensor $\sigma_{\alpha\beta}$ and the strain rate tensor $D_{\alpha\beta}$ for incompressible fluid is given by

$$\sigma_{\alpha\beta} = 2\mu D_{\alpha\beta} \tag{7}$$

The local viscosity μ is the a function of the invariants of the strain rate tensor. Here we use power law fluids, the viscosity is defined as

$$\mu = m\dot{\gamma}^{n-1} \tag{8}$$

where m and n are some constants, when $n > 1$ the fluid is categorized as shear-thickening fluid, when $0 < n < 1$ is categorized as shear-thinning fluid and when $n = 1$ the fluid is Newtonian. The local shear rate $\dot{\gamma}$ is then defined as

$$\dot{\gamma} = 2\sqrt{D_{\alpha\beta}D_{\alpha\beta}} \tag{9}$$

in LBM the strain rate tensor is given by

$$D_{\alpha\beta} = -\frac{3}{2\rho\tau}\sum_i (f_i - f_i^{eq})c_{i\alpha}c_{i\beta} \tag{10}$$

which is equal with equation 7.

For the validation study, we set the pressure gradient $\nabla P = G$ in the x-direction, and with power law parameters m and $n > 0$, in a channel width d, we will have the analytical solution of the velocity as

$$u_x(y) = \frac{n}{n+1}\left(\frac{G}{m}\right)^{1/n}\left(\left|\frac{d}{2}\right|^{(n+1)/n} - \left|\frac{d}{2} - y\right|^{(n+1)/n}\right) \tag{11}$$

3. Results and Discussion

3.1. Parallel plate validaton

The LBM was implemented on parallel plate for numerical validation. Simulation were run at low Reynold's number to ensure the flow is in laminar regime. The parameter n was varied to capture the shear thinning and thickening properties. We applied the bounceback rules for solid-fluid interaction, while Zou-He boundary conditions are applied for inlet and outlet flow [15]. The results displayed in Figure 1, it showed that the numerical solutions (marked) agreed very well with the analytical solutions (solid lines) Equation 11 for different n. We set $G = 1.0 \times 10^{-6}$ and $m = 0.02$ for validation study.

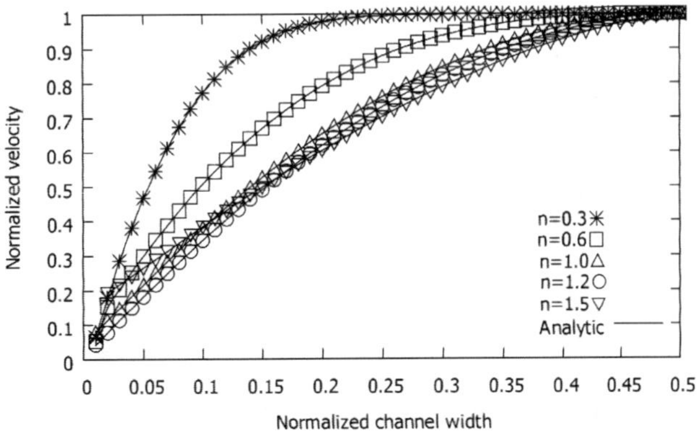

Figure 1. Velocity profile for flow in parallel plate for different n

3.2. Flow in fracture network

Our simulation are performed on the two-dimensional fracture network having size 342×680 in lattice unit. The fracture network geometry is developed using fBM method [7]. In this study we set the surface roughness $H = 0.8$ which is illustrated in Figure 2 and Figure 3. The fracture network was developed in such a way that they have same mean apertures for each branches. The power law fluid parameters are set with three different n (0.6; 1.0 1.5). Additional layer was added in inlet and outlet in order to perform periodical boundary conditions. The steady state conditions are reached after 50 000 iterations.

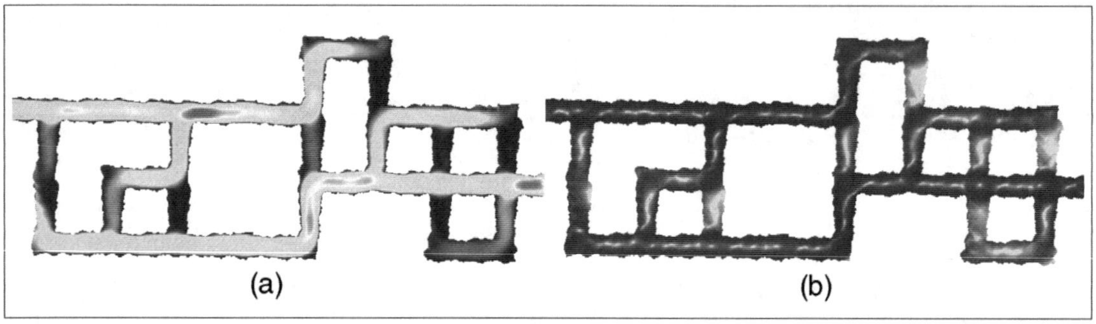

Figure 2. Velocity field (a) and viscosity profile (b) for fluids flowing through the fracture network with $n = 0.6$

Figure 3. Velocity field (a) and viscosity profile (b) for fluids flowing through the fracture network with $n = 1.5$

Figure 2 presents velocity and viscosity distribution for shear thinning ($n = 0.6$) fluid. The velocity profile has power-law fluid with viscosity, i.e. the higher shear rate, the lower viscosity. The red colour indicates the higher value while the blue is lower, this means the fluid will preferably flow in non-blue regions. In shear thickening fluid, the relationship between the velocity and viscosity is vice versa with the previous case as shown in Figure 3. This corresponds very well with the relationships of power law fluid. We then calculate the fluid flux for different constant pressure gradient. The slopes are tabulated in Table 1 and the plot is shown in Figure 4.

Table 1. Slopes from plots of $\log[u]$ vs. $\log[G]$

n	$1/n$	slope	% Error
0.6	1.6667	1.35576	0.18
1	1.0000	1.0000	0.00
1.5	0.6667	0.6667	0.00

Figure 4 shows the plot of $\log[u]$ versus $\log[G]$ for this simulation with different n. The error for Newtonian and shear thickening flow is less than 1%. Error increases with decreasing n or

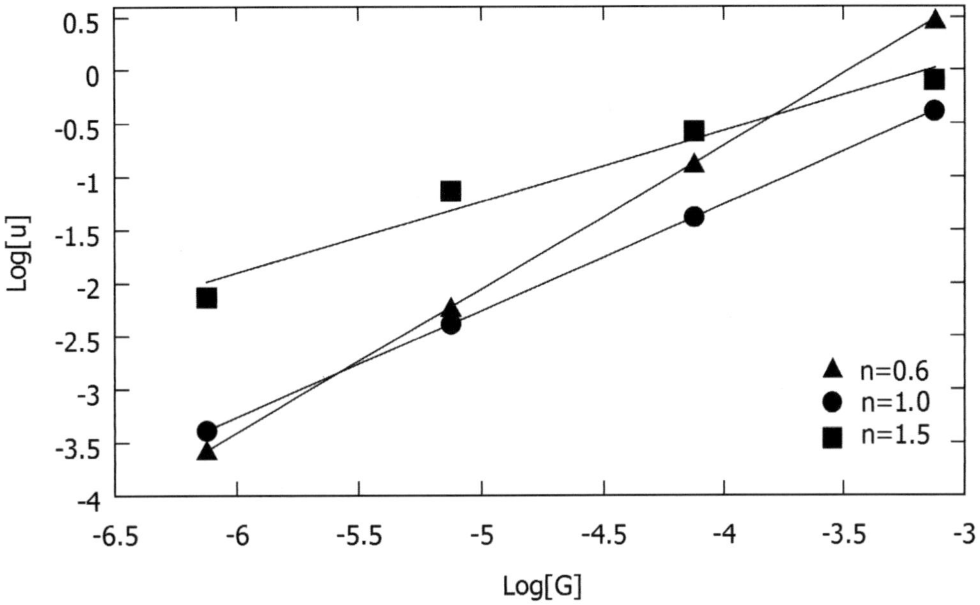

Figure 4. Logarithmic plot for the flow rate (log[u]) versus pressure difference (log[G]) for non-Newtonian flow with different n

fluid become more shear thinning. In power-law fluid model the problem may arise if the local strain-rate produces very small number or even zero, this only happen when n <1 because the local viscosity will grows up into very huge number and then we get inaccuracy results. On the other hand, the strain rate in shear thickening fluid is more distributed within the fracture and the local viscosity produces moderate values.

4. Conclusion

The power law non-Newtonian fluid flow in fracture network was studied numerically using lattice Boltzmann method. The flow rate through the fracture network is linear proportional to the logarithmic applied pressure difference. The Newtonian and shear thickening flow produced consistent results with theory, while for shear thinning flow the low shear rates contribute to the inaccuracy results.

Acknowledgment

The calculations were performed at the Guriang Computer Cluster http://guriang.unpad.ac.id. This work was financed by Directorate Degree for Higher Education (DIKTI) contract no 393/UN6.R/PL/2015.

References

[1] Yeo I W 2001. *Geosciences Journal* **5** 2.
[2] Dou Z and Zhou Z F 2014. *Water Science and Engineering* **7** 3.
[3] Hatiboglu C U and Tayfun 2007. *Phys. Rev. E* **76** 6.
[4] Shiyi C and Doolen G D 1998. *Annu. Rev. Fluid.* **30** 1.
[5] Succi S, Foti E and Higuera F 1989. *Europhys. Lett.* **10** 5.
[6] Mazzeo M D and Coveney P V 2008. *Comp. Phys. Comm.* **178** 12.
[7] Madadi M, VanSiclen V D and Sahimi M 2003. *Journal of Geophysical Research* **108** 2396.
[8] Aharonov E and Rothman D H 1993. *Geophys. Res. Lett* **20** 679.

[9] Boyd J, Buick J and Green S 2006. *J. Phys. A: Math. Gen.* **39** pp 14241-7.
[10] Drazer G and Koplik J 2000. *Phys. Rev. E.* **62** 6.
[11] Madadi M and Sahimi M 2003. *Phys. Rev. E* **67** 026309.
[12] Eker E and Akin S 2006. *Transport in Porous Media* **65**: pp 363-84.
[13] Chen S and Doolen G D 1988. *Ann. Rev. of Fluid Mechanics* **30** pp. 329-64.
[14] Bhatnagar P L, Gross E P and Krook M 1954. *Phys. Rev.* **511**.
[15] Zou Q and He X 1997. *Phys. Fluids* **9** 1591.